Volume 68

Socio-Genetics

Advances in Genetics, Volume 68

Serial Editors

Theodore Friedmann
University of California at San Diego, School of Medicine, USA
Jay C. Dunlap
Dartmouth Medical School, Hanover, NH, USA
Stephen F. Goodwin
University of Oxford, Oxford, UK

Volume 68

Socio-Genetics

Edited by

Marla B. Sokolowski

Department of Biology
University of Toronto, Mississauga
Ontario, Canada

AMSTERDAM • BOSTON • HEIDELBERG • LONDON
NEW YORK • OXFORD • PARIS • SAN DIEGO
SAN FRANCISCO • SINGAPORE • SYDNEY • TOKYO
Academic Press is an imprint of Elsevier

ELSEVIER

Academic Press is an imprint of Elsevier

525 B Street, Suite 1900, San Diego, CA 92101-4495, USA
30 Corporate Drive, Suite 400, Burlington, MA 01803, USA
32 Jamestown Road, London, NW1 7BY, UK
Radarweg 29, POBox 211, 1000 AE Amsterdam, The Netherlands

First edition 2009

ISBN: 978-0-12-374896-6
ISSN: 0065-2660

For information on all Academic Press publications
visit our website at elsevierdirect.com

Printed and bound in USA

09 10 11 12   10 9 8 7 6 5 4 3 2 1

Working together to grow
libraries in developing countries

www.elsevier.com | www.bookaid.org | www.sabre.org

ELSEVIER   BOOK AID
             International   Sabre Foundation

Contents

4   Approaching the Genomics of Risk-Taking Behavior   83
Alison M. Bell

Color Plate Section at the end of the Book

Contributors

Numbers in parentheses indicate the pages on which the authors' contributions begin.

Evan L. Ardiel (1) Brain Research Centre, University of British Columbia, Vancouver, British Columbia, Canada V6T 2B5, and Department of Psychology, University of British Columbia, Vancouver, British Columbia, Canada V6T 1Z4

Frank W. Avila (23) Department of Molecular Biology and Genetics, 421 Biotechnology Building, Cornell University, Ithaca, New York 14853, USA

Alison M. Bell (83) School of Integrative Biology, University of Illinois, Urbana-Champaign, Urbana, Illinois 61801, USA

Clement Y. Chow (23) Department of Molecular Biology and Genetics, 421 Biotechnology Building, Cornell University, Ithaca, New York 14853, USA

Jeffrey M. Donlea (57) Department of Anatomy and Neurobiology, Washington University School of Medicine, Campus Box 8108, St. Louis, Missouri, USA

Brooke A. LaFlamme (23) Department of Molecular Biology and Genetics, 421 Biotechnology Building, Cornell University, Ithaca, New York 14853, USA

Catharine H. Rankin (1) Brain Research Centre, University of British Columbia, Vancouver, British Columbia, Canada V6T 2B5, and Department of Psychology, University of British Columbia, Vancouver, British Columbia, Canada V6T 1Z4

C. Dustin Rubinstein (23) Department of Molecular Biology and Genetics, 421 Biotechnology Building, Cornell University, Ithaca, New York 14853, USA

Paul J. Shaw (57) Department of Anatomy and Neurobiology, Washington University School of Medicine, Campus Box 8108, St. Louis, Missouri, USA

Laura K. Sirot (23) Department of Molecular Biology and Genetics, 421 Biotechnology Building, Cornell University, Ithaca, New York 14853, USA

Jessica L. Sitnik (23) Department of Molecular Biology and Genetics, 421 Biotechnology Building, Cornell University, Ithaca, New York 14853, USA

Mariana F. Wolfner (23) Department of Molecular Biology and Genetics, 421 Biotechnology Building, Cornell University, Ithaca, New York 14853, USA

Preface

Why are few genes known that influence individual differences in social behavior? The first answer to this question is that social behavior is arguably the most complex of behavioral phenotypes. It involves dynamic interactions between at least two individuals and the behaviors performed by the individuals in a group are interdependent. That is, group membership affects social interactions. Specifically, the behavior of one individual is conditional on the behavior of others in the group. Studies of gene–environment interdependencies on social behavior are important because many individuals live in social environments. However, the "environment" for social behavior is not always easily quantified. The critical factors in an animal's social environment are more difficult to define than say, for example, the rearing temperature or nutritional history of an animal. But notice that even these abiotic factors can interact with social behavior. Thus, it should be clear that the question of what genes and environments are important for the development and functioning of normal social behavior is a complex one.

Where might genes that affect social behavior act in the nervous system? Genes that affect social behavior could act at the level of sensory input, integration of internal and external stimuli, and/or at the output of social behaviors. They could also act, for example, to influence the pattern of neural connectivity. Little is known about how animals continuously sample the social environment and respond to it and how prior social deprivation and social experience affects individual differences in social behavior. This is currently being studied at the level of individual genes, genomes, epigenomes, and the behavior itself. How these genes, genomes, and epigenomes interact with the external and internal environment of the animal to generate individual differences in social behavior is also of great interest. Overall, the richness of these interactive social phenotypes has begun to unfold because both mechanistic and evolutionary analyses are being used hand in hand.

It is a challenge to develop animal models for human disorders, particularly in the case of human social disorders such as autism or schizophrenia. For example, in the case of schizophrenia, a schizophrenic mouse model cannot tell if the mouse is having auditory hallucinations and we currently do not have the knowledge or technology available to determine if this is so. Another approach to the question of human social disorders is to identify most of the genes that affect social behavior in relatively simple animal models and then use these genes as candidates for human studies. The conservation of a gene's behavioral function has been found for many behaviors including the *period* gene and its role in circadian rhythms, the *foraging* gene and its roles in food-related behaviors, and genes in the dopamine pathway known to be involved in the reward system of many animals. These results suggest that a comparative analysis of the genetics and genomics of social behavior in a wide range of animals should aid in identifying genes that contribute predispositions to human social disorders.

This volume provides a sampling of some of the new and elegant research being performed in the area of the genetics of social behavior. Additional volumes on this topic are warranted as research moves forward in this exciting new area of investigation.

The first chapter is from Catherine Rankin's lab and it reviews what is known about *C. elegans* social behavior including the development of social behavior and the importance of mechanosensory input and social aspects of aggregation during foraging. The importance of identified sensory neurons and the factors important to the input of abiotic and biotic stimuli important for social behavior can be readily investigated in *C. elegans* with all of its neurons identified and much of the circuitry investigated.

The second chapter is from Mariana Wolfner's lab. It is written as a one-act play with the female and male social molecules interacting as the main players. This elegant research using the fruit fly involves the roles of male accessory gland proteins in postmating and other behavioral responses in the female. This research stands as a paradigm for how social interactions between the sexes have evolved at the molecular level.

The third chapter from Paul Shaw's lab provides a review of how social experience affects brain development in vertebrates and flies and then moves onto discuss exciting research describing how social experience interacts with sleep. Just as in humans when flies do too much partying they require more sleep. We learn that the similarities between fly and vertebrate's sleep are quite remarkable.

Finally, the fourth chapter is from Alison Bell's lab. Bell addresses risk taking/impulsivity in stickleback fish using an evolutionary approach. She shows that natural variants in risk taking exist in these fish and suggests a genomic approach for further investigating this interesting and important phenotype. The chapter also discusses the idea that an individual's behavior can affect group dynamics, another important consideration for the genetics of social behavior.

This new research area is at an exciting moment because it draws on issues from so many different research traditions (evolution, ecology, psychology, genetics, development, neurobiology) and it includes many model organisms including some not represented here (bacteria, dictyostelium, beetles, frogs, birds, rats, and voles and primates, just to name a few). Stay tuned for more exciting research findings!

Marla B. Sokolowski
University of Toronto, Mississauga

1

C. elegans: Social Interactions in a "Nonsocial" Animal

Evan L. Ardiel[*,†] and Catharine H. Rankin[*,†]

*Brain Research Centre, University of British Columbia, Vancouver, British Columbia, Canada V6T 2B5
†Department of Psychology, University of British Columbia, Vancouver, British Columbia, Canada V6T 1Z4

ABSTRACT

As self-fertilizing nematodes, *Caenorhabditis elegans* do not normally come to mind when one thinks of social animals. However, their reproductive mode is optimized for rapid population growth, and although they do not form structured societies, conspecifics are an important source of sensory input. A pheromone signal underlies multiple complex behaviors, including diapause, male-mating, and aggregation. The use of C. *elegans* in sociogenetics research allows for the

0065-2660/09 $35.00
DOI: 10.1016/S0065-2660(09)68001-9

analysis of social interactions at the level of genes, circuits, and behaviors. This chapter describes natural polymorphisms in *mab-23*, *plg-1*, *npr-1*, and *glb-5* as they relate to two *C. elegans* social behaviors: male-mating and aggregation. © 2009, Elsevier Inc.

Although it does not form anything resembling a structured society, the power of *Caenorhabditis elegans* as a model organism makes it an important system in the study of sociogenetics. Some 40 years ago Sydney Brenner chose the worm as the best metazoan in which to investigate development and the nervous system. It was small (approximately 1 mm) with a short life cycle (<4 days) and could be easily cultivated in the laboratory; furthermore, its mode of reproduction was ideal for genetics—self-fertilizing hermaphrodites could be easily inbred or crossed with males. Morphologically, *C. elegans* is relatively simple and its development is highly deterministic. As a result, the complete cell lineage and neural wiring diagram could be worked out; each adult hermaphrodite has only 959 cells, 302 of which are neurons forming about 5000 chemical synapses, 600 gap junctions, and 2000 neuromuscular junctions (Sulston and Horvitz, 1977; Sulston *et al.*, 1983; White *et al.*, 1986). The worm's transparency grants easy access to cells for targeted laser ablation and *in vivo* imaging of fluorescent markers. The genome has now been mapped and sequenced and thousands of mutants and RNAi constructs are readily available. Clearly, a powerful model system, but is it social? Although dense populations are rarely found in nature (Barrière and Félix, 2005), the boom and bust strategy of resource depletion suggest that conspecifics are a particularly salient feature of the environment. Indeed, pheromones influence gene expression and two types of social behavior have been well documented: male-mating and aggregation.

Conspecifics can affect one another in a wide variety of ways, ranging from providing simple sensory input (i.e., visual, auditory, or mechanosensory stimulation) to more complex emotional stimulation (nurturing, pair-bonding, dominance, etc.). A constructive and parsimonious approach to study "social" behavior is to ascertain what level/type of stimulation is required from conspecifics and to determine whether some other type of stimulation, rather than a conspecific, can have the same effect. To say this in a different way, the issue is whether the effect/behavior is mediated by a specific drive to interact with a conspecific or whether the conspecific produces some critical/important stimulus conditions that do not actually require conspecifics. A useful approach to studying the importance of a particular sensory input is to remove it and see what happens. For example, Rose *et al.* (2005) removed all conspecific cues by raising worms in isolation and compared these isolate-reared worms to group reared worms on several measures. They found that worms reared in social isolation were less responsive to mechanical stimulation (Fig. 1.1A) and had weaker synaptic connections in the underlying neural circuit, as compared to

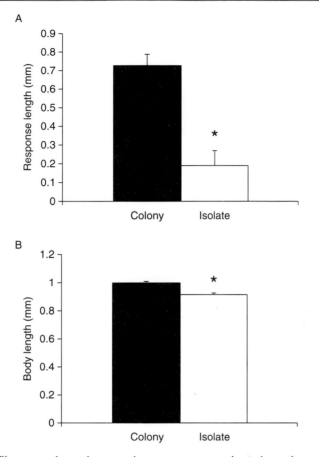

Figure 1.1. Worms reared in isolation are less responsive to mechanical stimulation than worms
reared in colonies (A). Responsiveness is scored as the distance a worm swims backwards
following a tap to the side of its Petri plate. Worms reared in isolation are shorter than
worms reared in colonies (B). *$p < 0.05$.

worms reared in colonies of 30–40. These differences were not the result of social
deprivation *per se*, as they could be reversed during development by simply
dropping the Petri plate containing an isolated worm (Rai and Rankin, 2007;
Rose *et al.*, 2005). Thus, conspecifics appeared to act as a source of mechanosensory stimulation during the maturation of the neural circuit. The effects of
isolation on the mechanosensory response were mediated by a gene, *glr-1*, which
encodes a glutamate receptor subunit. Rose *et al.* (2005) also noted that isolate-
reared worms were smaller (Fig. 1.1B) and had a delayed onset of egg-laying
compared to colony worms. No amount of mechanical stimulation could reverse
this effect, but isolated worms did grow larger if they were transferred into

colonies before the third larval molt (Rai and Rankin, 2007). This suggests that social interactions with other worms can influence adult body size and that there is a critical period for them to do so. Worms with mutations in *glr-1* did show the same effect of isolation on body size as wild-type worms, suggesting that this gene pathway is not involved in the effects of isolation on body size. The current work is attempting to identify the relevant conspecific cues that influence adult body size, the neurons through which these cues are registered, and the genes required to process them. Thus, the effects of isolation on mechanosensory behavior are mediated by simple mechanical stimulation-activating *glr-1* receptors; in contrast the effects of isolation on body size appear to be mediated by the physical presence of conspecifics and do not involve *glr-1* receptors.

At the other end of the population density continuum, social interactions are also important as populations become overcrowded. Under optimal conditions, *C. elegans* develops through four larval stages (L1–L4) to egg-laying adult in about 3 days. However, high population density and limited food promote a diversion into a larval diapause state known as dauer, which can last up to 4 months. Population density is assessed indirectly by the concentration of dauer pheromone, which is composed of several structurally related ascarosides (derivatives of the dideoxysugar ascarylose; Butcher *et al.*, 2007) constitutively secreted by worms (see Edison (2009) for an excellent review on *C. elegans* pheromones). Mutant analysis has identified at least four evolutionarily conserved signal transduction pathways regulating dauer formation. They are a guanylyl cyclase pathway, a TGFβ-like pathway, an insulin-like pathway, and a steroid hormone pathway. Morphologically specialized for long-term survival and dispersal, dauer larvae appear thin and dense and exhibit distinctly different behavioral patterns from developing larvae. Pharyngeal pumping is suppressed (Cassada and Russell, 1975) and movement is limited, although they do respond to touch. Dauer larvae may also climb objects and wave their body in the air, a behavior likely leading to insect-mediated dispersal. There is considerable reorganization of the nervous system, as several neurons adopt dauer-specific morphologies and positions. As would be expected, metabolism also changes to meet the demands of long-term survival in the absence of food, and the mouth is closed off. Altering another's life history is the most profound effect that could arise from a social interaction.

Robinson *et al.* (2008) described two key "vectors of influence" linking genes, neural circuits and social behaviors. The first is the effects of social interactions on brain gene expression and behavior. As highlighted above, dauer pheromone can have a profound impact on gene expression, brain function, and behavior. The second key vector of influence is genetic variation on social behaviors. This is the focus of the remainder of our chapter as we discuss research on two behaviors: male-mating and aggregation.

I. MALE-MATING

The interaction of a male and female during sexual reproduction is one of the most ancient of social behaviors. Although derived from an obligate-outcrossing, male–female ancestor, there are no female C. *elegans*, just males and hermaphrodites. Hermaphrodites produce a limiting amount of sperm and a large number of oocytes. The sperm fertilizes the eggs with nearly 100% efficiency, so a hermaphrodite can sire about 300 self-progeny, but they can also mate with males to sire over 1000 cross-progeny. Copulation is arguably the most complex social behavior of the worm, although the hermaphrodites play a mostly passive role.

Male-mating comprises several stereotyped sub-behaviors (Fig. 1.2; Liu and Sternberg, 1995). Males must find a partner, respond to contact, arch around the head or tail, locate the vulva, insert their spicules, and ejaculate (see Fig. 1.2). Step 1: *Hermaphrodite localization*. Males are attracted to a chemical cue secreted by hermaphrodites (Simon and Sternberg, 2002). Sensed by the ASK and CEM sensory neurons, the cue is composed of a synergistic blend of dauer-inducing ascarosides (Macosko et al., 2009; Srinivasan et al., 2008). Step 2: *Response to contact*. Upon contact, the male stops, presses the ventral side of its tail against its partner and begins swimming backward. This step is mediated by the sensory rays in the male's tail, likely responding to both chemo- and mechanosensory cues (Barr and Sternberg, 1999; Liu and Sternberg, 1995). Each of the nine bilaterally symmetrical rays is composed of a structural cell and two sensory neurons, RnA and RnB, where n = ray number, with 1 being the most anterior and 9 being the most posterior. Step 3: *Turning*. If the male does not encounter the vulva before reaching the end of the hermaphrodite (i.e., it's on the other side), he initiates a sharp ventral turn around the head or tail. Turning is mediated by the dopaminergic R7A and R9A ray neurons and the serotonergic R9B and CP neurons (Liu and Sternberg, 1995; Loer and Kenyon, 1993). The male-specific CP motor neurons synapse onto the diagonal muscles controlling ventral/dorsal flexion of the tail (White, 1986). Step 4: *Vulva localization*. Males continue to swim backward with their tail pressed against the hermaphrodite until they encounter its vulva. Usually they stop on their first encounter, but if they miss it they will keep searching around and around. Once the vulva is located, the male gets in position with a slow back and forth motion. Proper execution of this step requires the hook, the postcloacal sensilla, and the spicules (Liu and Sternberg, 1995). Step 5: *Spicule insertion*. During vulval contact, the protractor muscles repeatedly contract, causing the tips of the spicules to prod at the vulva. Once the spicules partially penetrate, the protractors completely contract so as to extend the spicules through the vulva, at which point all spicule movements cease. Step 6: *Ejaculation*. Over the course of about 4 s, 30–180 sperms are released from the seminal vesicle. They move through the vas deferens and cloaca and into the vulva. The SPV sensory neuron

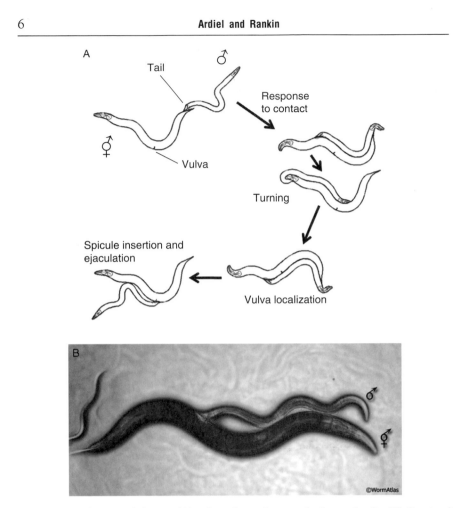

Figure 1.2. Male-mating behaviors (A) and a male copulating with a hermaphrodite (B). Reprinted with permission from www.wormatlas.org.

in the spicule plays an important role in the regulation of this process. If they are ablated, worms ejaculate prematurely (Liu and Sternberg, 1995). The spicules are retracted about a minute after ejaculation and the couple remains in contact for a few seconds.

Many molecules regulating male-mating have been identified through mutagenesis screens and mutant analysis of candidate genes. Most are required for cilia development, neurotransmitter biosynthesis and transmission, or male tail morphogenesis. Of particular interest for this chapter are natural polymorphisms in genes influencing social behaviors. So far two have been identified for male-mating.

A. *mab*-23

Hodgkin and Doniach (1997) identified male infertility in a strain that had been isolated in Vancouver, British Columbia. The source of the impotency was determined to be a recessive allele of *mab-23*, which results in a grossly misshapen male tail. In terms of male-mating sub-behaviors, these males are defective in positioning their tail and arching around the tip of the hermaphrodite, as a result, they lose contact with their partner before locating the vulva (Lints and Emmons, 2002). Lints and Emmons (2002) found that *mab-23* encoded a DM domain transcription factor required for the patterning of ray neurons, differentiation of male sex muscles, and morphogenesis of the posterior hypodermis, spicules, and proctodeum. DM domain genes are thought to be derived from an ancestral male sexual regulator. Although *mab-23* is expressed in both males and hermaphrodites in both sex-specific and nonspecific tissues, its loss only appears to affect males. An EMS-induced mutation and RNAi knockdown of *mab-23* resulted in the same defects as in the Vancouver strain, which has a phenylalanine replacing a conserved cysteine in the DNA-binding domain (Lints and Emmons, 2002). A naturally occurring *mab-23* loss-of-function mutation suggests that males are dispensable for survival of the species, at least in Vancouver.

B. *plg*-1

During the postcoital phase, males of some strains cover the hermaphrodite's vulva with a gelatinous mass. Although males can still inseminate plugged hermaphrodites, the copulatory plug lessens the likelihood of a second mating by lengthening the time it takes a male to locate the vulva and insert its spicules (Barker, 1994; Hodgkin and Doniach, 1997). Other *Caenorhabditis* species also plug the vulva after copulation, suggesting that it is an ancestral trait for mate guarding. The laboratory wild-type strain, N2, does not deposit copulatory plugs, nor do multiple strains isolated in Europe, North America, and Australia. Hodgkin and Doniach (1997) determined that plug production could be attributed to a dominant allele at a single locus on chromosome III, *plg-1*. Palopoli *et al.* (2008) found *plg-1* to encode a novel mucin-like gene strongly expressed in 12 sperm-storing cells of the vas deferens. Mucins are large glycoproteins and PLG-1 is predicted to be a major component of the gelatinous blob. All non-plugging strains contained the same retrotransposon insertion in *plg-1* and no stable mRNA product could be detected (Palopoli *et al.*, 2008). Transgenic expression of a *plg-1* clone without the retrotransposon was sufficient to induce plug deposition (Palopoli *et al.*, 2008). Although the *plg-1* locus has no direct influence on the behavior of the mating male, it results in a copulatory plug in the hermaphrodite vulva which does influence male-mating behavior.

Males arise through rare instances of X chromosome loss during hermaphroditic gametogenesis and contribute at a low rate of outcrossing thereafter. The appearance of hermaphroditism in C. *elegans* likely reduced selective pressure on genes for male–male competition, allowing for the appearance and proliferation of the *plg-1* loss-of-function allele.

C. Other natural variation in male-mating

Although the underlying genetic components are not fully understood, researchers have documented other natural variation in C. *elegans* mating. The first relates to virility (total lifetime male fertility). Wild-type males do not mate successfully after 4 days of adulthood at 20 °C and never sire more than 3000 progeny. In contrast, a strain from a garden compost heap in Palo Alto, California, was found to mate for up to 12 days and sire over 9000 progeny (Hodgkin and Doniach, 1997). Another example of increased male-mating comes from wild-isolate strains from Hawaii, USA and Beauchene, France, both of which were shown to dramatically increase outcrossing rates following passage through the dauer stage, presumably as a means to increase genetic diversity during times of stress (Morran *et al.*, 2009). Worms of the standard wild-type (N2) population showed no such increase. Finally, there is evidence for homosexual behavior in a strain of worms from Adelaide, Australia. Single-sex groups of males typically congregate into clumps of animals attempting to mate with one another. However, males of the Australian strain actually ejaculate, depositing copulatory plugs over the excretory pore of other males (Gems and Riddle, 2000). Intriguingly, they are not generally attracted to males, as they do not plug the excretory pore of other strains (i.e., wild type).

The purpose of male-mating behavior fits our standard idea of a "social" behavior, its objective is propagation of the species, and it is preferentially aimed at conspecific hermaphrodites; however, in the absence of conspecific hermaphrodites, males will attempt to copulate with inappropriate targets (i.e., other males, other species, etc.). It requires specialized morphological features, complex behavioral patterns, and species-specific pheromone signals for successful mating to occur.

II. AGGREGATION

Aggregation of conspecifics is prevalent throughout the animal kingdom, from swarms of insects to schools of fish to herds of ungulates. It can greatly influence population dynamics, community structure, and biodiversity. Aggregation is common in nematodes, for example, the formation of *Nippostrongylus brasiliensis* colonies in the vertebrate gut (Roberts and Thorson, 1977) and the swarming of

Tylenchorhynchus martini during rapid host-plant growth (Hollis and McBride, 1962). It is not clear if aggregation reflects a specific drive to be with conspecifics or is a response to other stimulus conditions that correlate with high densities of conspecifics. Despite its prevalence, little is known about the genetic and neural basis of aggregation. *C. elegans* is an ideal system in which to begin the dissection.

A. *npr*-1

Under standard laboratory conditions, *C. elegans* wild isolates exhibit one of two foraging behaviors—"solitary" or "social." The laboratory wild-type strain, N2, is a solitary feeder, distributing evenly over a lawn of *Escherichia coli*, but clumping as the bacterial food source becomes limiting. Social strains clump when food is bountiful, accumulating where bacteria grows thickest, which is usually at the border of a bacterial lawn. This behavior is called "bordering." In the absence of food, both social and solitary strains disperse; however, social strains begin to clump following prolonged food deprivation (Rogers *et al.*, 2006). Unlike solitary strains, social strains maintain rapid locomotion in the presence of food (de Bono and Bargmann, 1998) and following starvation (Rogers *et al.*, 2006). Social feeders also have an increased tendency to burrow into the agar medium (Hodgkin and Doniach, 1997). It is important to note that the behaviors of social strains are not interdependent, that is, rapid locomotion alone is neither necessary nor sufficient for clumping or bordering—sluggish social mutants still clumped and bordered, but hyperactive solitary mutants did not (de Bono and Bargmann, 1998). Furthermore, social strains still clumped in the absence of borders (Rogers *et al.*, 2006) and still bordered and swam fast when reared alone (de Bono and Bargmann, 1998).

In a genetic screen of the N2 solitary strain, several mutants were identified that had all of the characteristics of a social strain, that is, clumping (Fig. 1.3), bordering, burrowing, and fast swimming. The alleles did not complement and all mapped to the same position on the X chromosome, suggesting that a mutation in a single gene could switch a solitary feeder into a social feeder. de Bono and Bargmann (1998) cloned the gene and found it to encode a protein they called NPR-1 (for neuropeptide receptor resemblance), a seven-transmembrane domain receptor most similar to members of the neuropeptide Y (NPY) receptor family. Compared to the wild social strains, the *npr-1* null mutants were hypersocial. To get at the allelic difference underlying the natural behavioral polymorphism, de Bono and Bargmann (1998) sequenced the *npr-1* open reading frame of 17 strains originally isolated from the wild in various parts of the world. They found that the natural variation in social behavior in all of the strains could be attributed to a single amino acid substitution. This sequencing showed that all five solitary strains had an alanine (V), where all 12 social strains had a phenylalanine (F)—position 215, which is located in the third intracellular

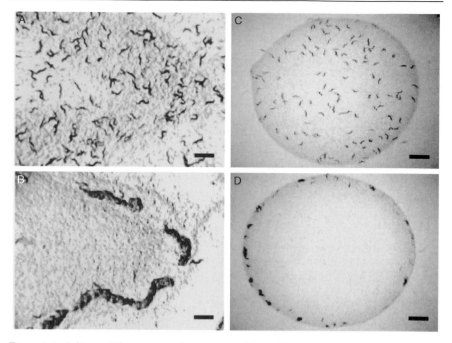

Figure 1.3. Solitary wild-type worms disperse over a lawn of bacteria (A and C) and social *npr-1* strains form clumps (B and D). The scale bars represent 1 mm in (A) and (B) and 2.5 mm in (C) and (D). Reprinted from de Bono and Bargmann (1998), with permission from Elsevier.

loop close to the fifth transmembrane domain. This region is important for G-protein coupling in many seven-transmembrane receptors. The NPR-1 amino acid sequence was otherwise identical among the strains, but other polymorphisms at the *npr-1* locus and throughout the genome confirmed that the worms were not just multiple isolates of the same strain. Transgenic expression of *npr-1* 215V in either wild social strains or *npr-1* null mutants induced solitary behavior (de Bono and Bargmann, 1998). Overexpression of *npr-1* 215F also induced solitary behavior in *npr-1* null mutants (de Bono and Bargmann, 1998). Therefore, both NPR-1 215V and NPR-1 215F repress social behavior, but NPR-1 215V is more potent. It is important to emphasize that the environment–genome interaction ultimately dictates behavior. Therefore, labeling strains as solitary or social is only instructive *under standard laboratory conditions*. Even "solitary" strains will aggregate as food becomes limiting or following an ethanol exposure (Davies *et al.*, 2004).

Evidence suggests that the *npr-1* 215V allele found in solitary strains is evolutionarily derived. Rogers *et al.* (2003) sequenced putative NPR-1 orthologs from three other *Caenorhabditis* species: *C. briggsae*, *C. remanei*, and *C. brenneri*.

Each had phenylalanine at the position corresponding to NPR-1 residue 215, suggesting that *npr-1* 215V arose as a gain-of-function allele from the more ancestral *npr-1* 215F. Recently, McGrath *et al.* (2009) sequenced 203 *C. elegans* strains that had previously been isolated from Europe, North America, Africa, South America, Australia, and Japan and cultured in laboratories for many years. Only 12 had an alanine at NPR-1 residue 215. It may be that the environmental conditions favoring solitary behavior occurred more rarely or were less likely to be sampled by those hunting *C. elegans*. In an attempt to isolate more solitary strains from the wild, McGrath *et al.* (2009) turned to Pasadena, California, where 9 of the 12 solitary strains originated. Of 55 newly isolated strains sampled here, none had *npr-1* 215V. An intriguing hypothesis for the rare occurrence of this allele is that it may have arisen in the laboratory (see McGrath *et al.*, 2009). After all, 11 of the 12 solitary strains had been in laboratory stocks for over 15 years. The laboratory wild-type strain, N2, was isolated around 1951 by L. N. Staniland from mushroom compost near Bristol, England. The earliest frozen cultures are from 1969. It is possible that the solitary feeding phenotype of N2 arose spontaneously and was selected for under laboratory conditions, where solitary strains spend more time on good food sources and are less likely to be killed by pathogenic bacteria (Gloria-Soria and Azevedo, 2008; Reddy *et al.*, 2009; Styer *et al.*, 2008). Researchers could have also inadvertently selected for *npr-1* 215V by picking single animals from the agar surface and avoiding the clumping and burrowing social worms. Further sampling and sequencing will reveal how often (if ever) the 215V allele occurs in nature.

B. NPR-1 ligands

Although NPR-1 is most similar to members of the NPY receptor family, there are no obvious NPY homologs in the genome of *C. elegans*. There is however evidence that NPY is evolutionarily related to invertebrate FMRFamide-like peptides (FLPs; Rajpara *et al.*, 1992; Tensen *et al.*, 1998). Furthermore, FLPs have been linked to NPR-1-like receptors in *Drosophila* and humans (Feng *et al.*, 2003; Hinuma *et al.*, 2000). *C. elegans* has well over 60 predicted FLPs (some of which have been detected biochemically; Husson *et al.*, 2005) encoded by 31 genes (Husson and Schoofs, 2007; Kim and Li, 2004; Li *et al.*, 1998; McVeigh *et al.*, 2005). To identify its ligand, Kubiak *et al.* (2003) expressed NPR-1 in Chinese hamster ovary (CHO) cells in culture and used the $[^{35}S]GTP\gamma$ binding assay to test about 200 synthetic FLPs from *C. elegans* and other invertebrates. A single active peptide was identified: GLGPRPLRFamide. First isolated in the parasitic nematode *Ascaris suum* (Cowden and Stretton, 1995), GLGPRPLRFamide is the sole product of *flp-21* in *C. elegans*. Using the CHO cells, Kubiak *et al.* (2003) found that NPR-1 215V had higher binding and functional activity than NPR-1 215F, which is consistent with the predictions made by genetic analysis.

Rogers *et al.* (2003) tested the ability of C. *elegans* FLPs to stimulate NPR-1 in *Xenopus laevis* oocytes. They too identified the *flp-21* ligand as an activator of NPR-1. Consistent with the findings of Kubiak *et al.* (2003), dose–response curves revealed that NPR-1 215V signaled more strongly than NPR-1 215F at subsaturating concentrations of GLGPRPLRFamide. Furthermore, Rogers *et al.* (2003) identified peptides that only activated the 215V form of NPR-1 in X. *laevis* oocytes. They were six predicted FLPs encoded by *flp-18*. Using the C. *elegans* pharynx as an expression system, they again searched for activators of NPR-1, testing a total of 59 FLPs encoded by 22 genes. Only *flp-21* and *flp-18* ligands activated NPR-1, causing a decrease in the frequency of pharyngeal action potentials at picomolar concentrations. In contrast to their previous finding, *flp-18* ligands activated both forms of NPR-1 in the pharynx, suggesting the signaling capabilities of NPR-1 215V and NPR-1 215F depend upon their cellular context. These studies demonstrate that *flp-18* and *-21* encode cognate ligands for NPR-1. They also support the early genetic data, that is, both forms of NPR-1 are functional, but NPR-1 215V is more potent.

If this hypothesis is correct, NPR-1 ligands should function to inhibit social behavior. Consistent with this role, overexpression of the FLP-21 ligand (by either high-copy gene insertion or injection of the peptide) suppressed social feeding in *npr-1* 215F animals, but had no effect on the aggregating behavior of *npr-1* null mutants (Rogers *et al.*, 2003). Furthermore, a loss-of-function mutation in *flp-21* slightly increased aggregation in both *npr-1* 215V and *npr-1* 215F animals (Rogers *et al.*, 2003). Activity of other NPR-1 ligands, that is, those encoded by *flp-18*, could explain why *flp-21* mutants did not aggregate as strongly as *npr-1* null mutants. Overexpression of FLP-18 impaired locomotion, precluding behavioral analysis. To identify the expression pattern of the NPR-1 ligands, Rogers *et al.* (2003) fused GFP to the promoter regions of *flp-18* and *flp-21*. Both genes were expressed in a small number of cells, *flp-18* in interneurons AVA, AIY, and RIG, motor neuron RIM, and pharyngeal neurons M2 and M3 and *flp-21* in the intestine, as well as sensory neurons ADL, ASH, and ASE, motor neuron URA, and pharyngeal neurons MC, M2, and M4.

C. NPR-1 neurophysiology

To determine where NPR-1 was expressed, Coates and de Bono (2002) imaged transgenic worms in which NPR-1 215V was fused to GFP. They saw NPR-1:: GFP fluorescence mainly in neurons, but also in a muscle in the terminal bulb of the pharynx, and sometimes in the excretory duct cell and excretory canal. In the nervous system, NPR-1 expressing cells included sensory neurons AQR, ASE, ASG, ASH, URX, IL2, OLQ, PQR, PHA, and PHB, interneurons AUA, SAAD, RIV, RIG, and SDQ, and motor neurons RMG, SMBD, VD, and DD. Consistent with its role as a neuropeptide receptor, NPR-1::GFP localized to cell bodies, axons, and dendrites and was present throughout postembryonic development. To determine when NPR-1 was functioning, Coates and de Bono (2002) used transgenic *npr-1* null mutants in which NPR-1 215V was under the control of the *hsp-16* heat-inducible promoter. These social mutants displayed reduced clumping and bordering within 60-min of hot-shock-induced NPR-1 expression, suggesting that NPR-1 activity suppresses social behavior acutely, as opposed to regulating neurodevelopment.

Recently, Macosko *et al.* (2009) proposed a hub-and-spoke neural circuit for the integration of sensory cues and regulation of social behaviors by NPR-1. In an attempt to identify where NPR-1 215V was functioning to inhibit social behaviors, they used various promoters to rescue its expression in subsets of neurons. They found that RMG expression of *npr-1* was necessary for robust suppression of the aggregation, bordering, and rapid locomotion phenotypes of *npr-1* null mutants. Coates and de Bono (2002) had previously reported partial suppression of aggregation and bordering by driving a genomic *npr-1* clone with *gcy-32* or *flp-8* promoters, which overlap only in oxygen-sensing URX neurons. Macosko *et al.* (2009) got no rescue using the same promoters to drive *npr-1* cDNA. The use of different rescue constructs likely explains this discrepancy. One possibility is that the genomic clone mediated higher *npr-1* expression in URX or perhaps the introns were acting as tissue-specific or nonspecific transcriptional enhancers, leading to RMG expression of *npr-1* regardless of promoter (Macosko *et al.*, 2009). There are no known RMG specific promoters, so to determine if NPR-1 in RMG was sufficient to mediate social behaviors, Macosko *et al.* (2009) used Cre–Lox recombination to express *npr-1* in only RMG (it was also actually expressed in M2, but this is a pharyngeal motor neuron synaptically isolated from neurons involved in social behaviors). Remarkably, *npr-1* expression in RMG was sufficient for the suppression of all social behaviors. An interneuron electrically coupled to six sensory neurons (White *et al.*, 1986), RMG is at the center of a gap-junction network for the integration of environmental information (Fig. 1.4).

Among the neurons coupled to RMG are the oxygen-sensing URX neuron and the nociceptive neurons ASH and ADL, all three of which had been previously shown to stimulate social behavior (Coates and de Bono, 2002; de Bono *et al.*, 2002). The role of URX in social behaviors will be discussed later in the chapter. Testing several sensory transduction mutants, de Bono *et al.* (2002) found that mutations in cation channel subunits of the temperature-sensitive transient receptor potential vanilloid (TRPV) type, *osm-9* and *ocr-2*, suppressed aggregation of *npr-1* null mutants. These two subunits are normally coexpressed in six neurons, experiments using cell-specific rescue demonstrated that OCR-2 was required in either ASH or ADL to mediate aggregation. The novel single-pass transmembrane protein, ODR-4, was also required in ADL (de Bono *et al.*, 2002) for aggregation to occur. To confirm their role

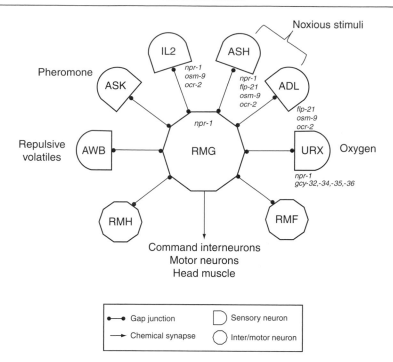

Figure 1.4. Hub-and-spoke circuit of neurons electrically coupled to RMG, including stimuli
detected by sensory neurons and expression pattern of several relevant genes, that is,
npr-1, flp-21, osm-9, ocr-2, and *gcy-32, -34, -35, -36.* Adapted from Macosko *et al.*
(2009).

in aggregation behavior, de Bono *et al.* (2002) ablated ASH and/or ADL and
found that simultaneous loss of these cells transformed social feeders into solitary
feeders. The ASH and ADL neurons mediate the avoidance response from a
variety of aversive stimuli, including hyperosmotic solutions and toxic heavy
metal ions (Bargmann *et al.*, 1990, Sambongi *et al.*, 1999). Consistent with these
observations, aversive stimuli may also promote group feeding. It may be that
under standard laboratory conditions, the moderately pathogenic bacteria food
source (*E. coli*) is sufficiently noxious to induce aggregation in NPR-1 215F
animals. In accordance with this hypothesis, *npr-1* null mutants actually dis-
persed on a lawn of harmless dead bacteria, but aggregated when exposed to the
odor of living *E. coli* (de Bono *et al.*, 2002), which has been shown to kill
C. elegans when the *E. coli* was grown on brain heart infusion agar medium
rather than on standard nematode growth medium agar (Garsin *et al.*, 2001).
In addition to URX, ASH, and ADL, RMG is coupled to AWB, which senses

repulsive odorants (Troemel *et al.*, 1997), ASK, which senses pheromones (Macosko *et al.*, 2009), IL2, which is a putative chemosensory neuron, and RMH and possibly RMF, which are both motor neurons.

Laser ablation of RMG had no obvious effect on solitary *npr-1* 215V animals, but suppressed the aggregation, bordering, and rapid locomotion of *npr-1* null mutants, suggesting that NPR-1 activity inhibits social behaviors by inhibiting RMG activity. How then does the circuit function? RMG could be mediating aggregation via its own chemical synapses (to head muscles and interneurons that control locomotion) or by modulating the synaptic output of the cells to which it is electrically coupled. The answer appears to be that both of these are true. Macosko *et al.* (2009) inhibited synaptic transmission of various neurons in the hub-and-spoke circuit by selective cell-specific expression of the light chain of tetanus toxin that cleaves synaptobrevin and blocks vesicle fusion, thereby blocking neurotransmitter release from the neuron in which it is expressed. They found that synaptic silencing of RMG partially suppressed the aggregating behavior of the *npr-1* null mutants, as did simultaneous silencing of ASK and ASJ. These effects were additive, that is, silencing RMG, ASK, and ASJ further suppressed aggregation, indicating that synaptic outputs for aggregation are distributed.

ASK sensory neurons were of particular interest because of their role in mediating male attraction to the ascaroside compounds released by hermaphrodites (Srinivasan *et al.*, 2008). Using the genetically encoded calcium indicator, G-CaMP, Macosko *et al.* (2009) showed that ASK hyperpolarized in response to ascaroside cocktails. They went on to demonstrate that the signal propagated to AIY interneurons via inhibitory synapses and that the magnitude of the ASK and AIY response was amplified in an *npr-1* null mutant background. The amplification of this response appears to be behaviorally relevant, as *npr-1* null mutants were actually attracted to the low levels of ascarosides that repelled the solitary wild-type strain (Macosko *et al.*, 2009). Expressing *npr-1* 215V in RMG was sufficient to induce ascaroside repulsion in *npr-1* null mutants. These data suggest that RMG stimulates pheromone attraction by enhancing sensory signaling in ASK. Therefore, a component of aggregation behavior is directed movement toward other worms, the point source of secreted ascarosides. Importantly, mutations in *daf-22* (which is required for the biosynthesis of ascarosides; Butcher *et al.*, 2009; Golden and Riddle, 1985) did not suppress the clumping behavior of *npr-1* null mutants (de Bono *et al.*, 2002). Therefore, attractive pheromones cannot be the only thing bringing social worms together. Indeed a shared preference for low oxygen levels has been identified as a second component of *npr-1*-mediated aggregation.

D. Oxygen-mediated aggregation

In a genetic screen for suppressors of npr-1, Cheung et al. (2004) identified multiple alleles of gcy-35 and gcy-36, which encode soluble guanylate cyclases (sGCs). Null mutations in either gene abolished aggregation and reduced bordering, but did not suppress rapid locomotion. Aggregation and bordering of npr-1 mutants was also suppressed by simultaneous disruption of two other sGCs, GCY-32 and GCY-34 (Cheung et al., 2005). Previously characterized sGCs were all activated by nitric oxide (NO), but this was not the predicted ligand for sGCs in C. elegans because the worm genome does not encode NO synthase (the NO-producing enzyme). Oxygen was proposed as a candidate ligand, as it binds other hemoproteins. Gray et al. (2004) tested the aerotaxis behavior of gcy-35 mutants. They found them to be defective, failing to avoid hyperoxia in gas-phase oxygen gradients. They went on to show that the GCY-35 heme domain binds molecular oxygen, implicating it as an oxygen sensor. GCY-36, -32, and -34 are predicted to have similar or related functions in oxygen sensation. Expressed in AQR, PQR, URX, ALN, PLN, PLM, BDU, SDQ, and AVM neurons, gcy-35 expression in AQR, PQR, and URX neurons was sufficient for normal aerotaxis and aggregation (Cheung et al., 2004; Gray et al., 2004). Previously, Coates and de Bono (2002) had found that disrupting the activity of these neurons (with a gain-of-function K^+ channel encoded by egl-2) suppressed aggregation of npr-1 null mutants. Taken together, these studies suggested that oxygen regulates social feeding through sGCs in URX, AQR, and PQR neurons. Testing the role of oxygen directly, Gray et al. (2004) demonstrated that the aggregation and bordering behavior of social strains was suppressed as ambient oxygen tensions dropped from 21% to 7% O_2.

Living at the air/water interface of rotting organic matter, C. elegans experience oxygen tensions ranging from 0% to 21% over short space and time scales. Oxygen rapidly diffuses through the worm and does not affect respiration until tensions fall below 4%. Worms can survive anoxia by entering suspended animation, but tensions between 0.01% and 0.1% are lethal (Nystul and Roth, 2004; Shen and Powell-Coffman, 2003). Excess oxygen is also harmful because of the production of toxic reactive oxygen species. In the absence of food, both social and solitary strains avoided hypoxia and hyperoxia in gradients of 0–21% O_2, preferring oxygen tensions around 7% (Gray et al., 2004). Aerotaxis to intermediate oxygen concentrations may be a means to stay below the Earth's surface, to prevent oxidative damage, or to locate oxygen-consuming microbial food. In the presence of food, social strains and npr-1 null mutants continued to accumulate around 7% O_2, while the solitary wild-type strain avoided hypoxia, but otherwise ignored the oxygen gradient (Gray et al., 2004). A similar phenomenon was observed by Cheung et al. (2005). Tracking locomotion within a food patch, they found that wild social strains and *npr-1* null mutants suppressed locomotion as oxygen tensions dropped from 21% to 11%, while oxygen tension had minimal effects on the slower moving solitary wild-type strain.

Gray *et al.* (2004) demonstrated that the aggregation and bordering behavior of social strains was suppressed as oxygen tensions dropped from 21% to the preferred 7%. Oxygen levels are not equivalent throughout bacterial lawns. Gray *et al.* (2004) found that the oxygen concentration at the border of the lawn was 13%, compared to 17% in the center. Thus, the bordering of social strains is likely mediated by hyperoxia avoidance, a behavior suppressed by solitary strains in the presence of food. Furthermore, aggregating worms locally deplete oxygen. Rogers *et al.* (2006) measured the oxygen concentration in a clump of worms and found it to be only 6%, as compared to 20% outside the clump. They proposed that aggregating worms stay in the clump by responding to this oxygen gradient. Monitoring behavior of clumping social strains and *npr-1* null mutants, they identified a stereotyped pattern. Aggregation began with two worms contacting along their body length and suppressing forward locomotion. They then swam back and forth to stay beside each other. Others joined the pair and a clump was formed. When the front of a worm stuck out of the clump, it either initiated a short reversal (1/6 of a body length) or turned to get back to the group. If its tail emerged from the clump, the worm initiated forward locomotion to get back in. They hypothesized that anterior and posterior oxygen sensors mediated behavioral responses to the sudden rise in oxygen concentration as the worm left a clump. The behaviors promote a return to the clump and preferred oxygen level.

Recently, McGrath *et al.* (2009) showed that natural variation in *glb-5* could modulate aggregation of social strains. GLB-5 is a hexacoordinated globin functioning in the AQR, PQR, and URX neurons to sensitize their oxygen response (McGrath *et al.*, 2009; Persson *et al.*, 2009). In wild-type worms, the genomic sequence of *glb-5* contains a 765-bp duplication/insertion not present in any "social" strains or other *Caenorhabditis* species (McGrath *et al.*, 2009; Persson *et al.*, 2009). The duplication results in an in-frame stop codon that truncates the last 179 amino acids (McGrath *et al.*, 2009). Despite its rarity in natural isolates, 11 of 12 "solitary" strains carry the *glb-5* duplication (McGrath *et al.*, 2009). The coinheritance pattern of *npr-1* and *glb-5* alleles suggests coselection and/or rare interbreeding of "social" and "solitary" strains. While GLB-5 activity did not influence the social behavior of the solitary NPR-1 215V animals, it suppressed the aggregation of "social" strains, as the percentage of animals in clumps was enhanced in worms homozygous for both the *npr-1* 215F allele and the duplicated *glb-5* allele (McGrath *et al.*, 2009).

E. Parallel pathway for aggregation

Although not as strong as in wild social strains, Thomas *et al.* (1993) observed aggregation and bordering in mutants for components of the TGF-β pathway that regulates dauer formation, for example the ligand, DAF-7. This pathway acts in parallel to NPR-1, as *daf-7;npr-1* double mutants had enhanced social behavior compared to either single mutant (de Bono *et al.*, 2002). Expressed in ASI chemosensory neurons, *daf-7* is transcriptionally repressed by crowding and induced by food (Ren *et al.*, 1996; Schackwitz *et al.*, 1996). As with *npr-1* mutants, *daf-7* mutants also exhibited robust hyperoxia avoidance in the presence or absence of food (Chang *et al.*, 2006), suggesting that like *npr-1* mutants, their propensity to clump at the border is mediated, at least in part, by a preference for lower oxygen concentrations.

    These studies offer interesting insights into the motivations driving aggregation in C. *elegans*. It appears as though lowered oxygen levels are a key characteristic of worm aggregations and may be sufficient to induce and maintain clumping. This suggests the hypothesis that it is not the company of other worms themselves that is motivating aggregation, it may just be the environmental conditions produced by aggregations that are attractive. However, the observation that individual worms are attracted to pheromonal cues of conspecifics suggests there might be additional motivation for aggregation beyond environmental conditions. Further studies on this will continue to enhance our understanding of why animals form social aggregations.

III. CONCLUSION

Despite the vast amount of information available on nearly every aspect of the worm's biology, remarkably little is known about its ecology. Its natural habitat is not certain, but it can be found in farmland and garden soil, compost heaps, and on a variety of carrier invertebrates, such as snails, slugs, and isopods, millipedes, and other arthropods. In the laboratory, it is reared on a smooth agar surface in a puddle of slightly pathogenic E. *coli*. As worms are studied in less artificial environments, new behaviors will undoubtedly be described. Furthermore, the majority of behavioral research to date has been conducted on a single genetic background. Looking at other members of the species is obviously necessary to appreciate natural variation in behaviors. One purpose of this chapter was to show that once natural variation is uncovered, C. *elegans* allows unprecedented access to the genes and neural circuits underlying the behavior. A second purpose was to dissect "social" behavior and examine motivations for behaviors considered to be "social" to ask whether the motivation was a specific drive to be with conspecifics or whether the presence of conspecifics generated stimulus conditions that then governed the behavior. Using C. *elegans* to approach social behavior from this reductionist perspective may provide insights into, and a new way to, dissect the evolution of more complex societies.

References

Bargmann, C. I., Thomas, J. H., and Horvitz, H. R. (1990). Chemosensory cell function in the behavior and development of *Caenorhabditis elegans*. *Cold Spring Harb. Symp. Quant. Biol.* **55**, 529–538.

Barker, D. M. (1994). Copulatory plugs and paternity assurance in the nematode *Caenorhabditis elegans*. *Anim. Behav.* **48**, 147–156.

Barr, M. M., and Sternberg, P. W. (1999). A polycystic kidney-disease gene homologue required for male mating behavior in C. *elegans*. *Nature* **401**(6751), 386–389.

Barrière, A., and Félix, M. A. (2005). High local genetic diversity and low outcrossing rate in *Caenorhabditis elegans* natural populations. *Curr. Biol.* **15**(13), 1176–1184.

Butcher, R. A., Fujita, M., Schroeder, F. C., and Clardy, J. (2007). Small-molecule pheromones that control dauer development in *Caenorhabditis elegans*. *Nat. Chem. Biol.* **3**(7), 420–422.

Butcher, R. A., Ragains, J. R., Li, W., Ruvkun, G., Clardy, J., and Mak, H. Y. (2009). Biosynthesis of the *Caenorhabditis elegans* dauer pheromone. *Proc. Natl. Acad. Sci. USA* **106**(6), 1875–1879.

Cassada, R. C., and Russell, R. L. (1975). The dauer larva, a post-embryonic developmental variant of the nematode *Caenorhabditis elegans*. *Dev. Biol.* **46**(2), 326–342.

Chang, A. J., Chronis, N., Karow, D. S., Marletta, M. A., and Bargmann, C. I. (2006). A distributed chemosensory circuit for oxygen preference in C. *elegans*. *PLoS Biol.* **4**(9), e274.

Cheung, B. H., Arellano-Carbajal, F., Rybicki, I., and de Bono, M. (2004). Soluble guanylate cyclases act in neurons exposed to the body fluid to promote C. *elegans* aggregation behavior. *Curr. Biol.* **14**(12), 1105–1111.

Cheung, B. H., Cohen, M., Rogers, C., Albayram, O., and de Bono, M. (2005). Experience-dependent modulation of C. *elegans* behavior by ambient oxygen. *Curr. Biol.* **15**(10), 905–917.

Coates, J. C., and de Bono, M. (2002). Antagonistic pathways in neurons exposed to body fluid regulate social feeding in *Caenorhabditis elegans*. *Nature* **419**(6910), 925–929.

Cowden, C., and Stretton, A. O. (1995). Eight novel FMRFamide-like neuropeptides isolated from the nematode *Ascaris suum*. *Peptides* **16**(3), 491–500.

Davies, A. G., Bettinger, J. C., Thiele, T. R., Judy, M. E., and McIntire, S. L. (2004). Natural variation in the *npr-1* gene modifies ethanol responses of wild strains of C. *elegans*. *Neuron* **42**(5), 731–743.

de Bono, M., and Bargmann, C. I. (1998). Natural variation in a neuropeptide Y receptor homolog modifies social behavior and food response in C. *elegans*. *Cell* **94**(5), 679–689.

de Bono, M., Tobin, D. M., Davis, M. W., Avery, L., and Bargmann, C. I. (2002). Social feeding in *Caenorhabditis elegans* is induced by neurons that detect aversive stimuli. *Nature* **419**(6910), 899–903.

Edison, A. S. (2009). *Caenorhabditis elegans* pheromones regulate multiple complex behaviors. *Curr. Opin. Neurobiol.* **19**(4), 378–388.

Feng, G., Reale, V., Chatwin, H., Kennedy, K., Venard, R., Ericsson, C., Yu, K., Evans, P. D., and Hall, L. M. (2003). Functional characterization of a neuropeptide F-like receptor from *Drosophila melanogaster*. *Eur. J. NeuroSci.* **18**(2), 227–238.

Garsin, D. A., Sifri, C. D., Mylonakis, E., Qin, X., Singh, K. V., Murray, B. E., Calderwood, S. B., and Ausubel, F. M. (2001). A simple model host for identifying gram-positive virulence factors. *Proc. Natl. Acad. Sci. USA* **98**(19), 10892–10897.

Gems, D., and Riddle, D. L. (2000). Genetic, behavioral and environmental determinants of male longevity in *Caenorhabditis elegans*. *Genetics* **154**(4), 1597–1610.

Gloria-Soria, A., and Azevedo, R. B. (2008). *npr-1* regulates foraging and dispersal strategies in *Caenorhabditis elegans*. *Curr. Biol.* **18**(21), 1694–1699.

Golden, J. W., and Riddle, D. L. (1985). A gene affecting production of the *Caenorhabditis elegans* dauer-inducing pheromone. *Mol. Gen. Genet.* **198**(3), 534–536.

Gray, J. M., Karow, D. S., Lu, H., Chang, A. J., Chang, J. S., Ellis, R. E., Marletta, M. A., and Bargmann, C. I. (2004). Oxygen sensation and social feeding mediated by a *C. elegans* guanylate cyclase homologue. *Nature* **430**(6997), 317–322.

Hinuma, S., Shintani, Y., Fukusumi, S., Iijima, N., Matsumoto, Y., Hosoya, M., Fujii, R., Watanabe, T., Kikuchi, K., Terao, Y., Yano, T., Yamamoto, T., et al. (2000). New neuropeptides containing carboxy-terminal RFamide and their receptor in mammals. *Nat. Cell Biol.* **2**(10), 703–708.

Hodgkin, J., and Doniach, T. (1997). Natural variation and copulatory plug formation in *Caenorhabditis elegans*. *Genetics* **146**(1), 149–164.

Hollis, J. P., and McBride, J. M. (1962). Induction of swarming in *Tylenchorhynchus martini* (Nematoda: *Tylenchida*). *Phytopathology* **52**, 14.

Husson, S. J., and Schoofs, L. (2007). Altered neuropeptide profile of *Caenorhabditis elegans* lacking the chaperone protein 7B2 as analyzed by mass spectrometry. *FEBS Lett.* **581**, 4288–4292.

Husson, S. J., Clynen, E., Baggerman, G., De Loof, A., and Schoofs, L. (2005). Discovering neuropeptides in *Caenorhabditis elegans* by two dimensional liquid chromatography and mass spectrometry. *Biochem. Biophys. Res. Commun.* **335**(1), 76–86.

Kim, K., and Li, C. (2004). Expression and regulation of an FMRFamide-related neuropeptide gene family in *Caenorhabditis elegans*. *J. Comp. Neurol.* **4754**(4), 540–550.

Kubiak, T. M., Larsen, M. J., Nulf, S. C., Zantello, M. R., Burton, K. J., Bowman, J. W., Modric, T., and Lowery, D. E. (2003). Differential activation of "social" and "solitary" variants of the *Caenorhabditis elegans* G protein-coupled receptor NPR-1 by its cognate ligand AF9. *J. Biol. Chem.* **278**(36), 33724–33729.

Li, C., Kim, K., and Nelson, L. S. (1998). FMRFamide-related neuropeptide gene family in *Caenorhabditis elegans*. *Brain Res.* **848**(1), 26–34.

Lints, R., and Emmons, S. W. (2002). Regulation of sex-specific differentiation and mating behavior in *C. elegans* by a new member of the DM domain transcription factor family. *Genes Dev.* **16**(18), 2390–2402.

Liu, K. S., and Sternberg, P. W. (1995). Sensory regulation of male mating behavior in *Caenorhabditis elegans*. *Neuron* **14**(1), 79–89.

Loer, C. M., and Kenyon, C. J. (1993). Serotonin-deficient mutants and male mating behavior in the nematode *Caenorhabditis elegans*. *J. Neurosci.* **13**(12), 5407–5417.

Macosko, E. Z., Pokala, N., Feinberg, E. H., Chalasani, S. H., Butcher, R. A., Clardy, J., and Bargmann, C. I. (2009). A hub-and-spoke circuit drives pheromone attraction and social behavior in *C. elegans*. *Nature* **458**(7242), 1171–1175.

McGrath, P. T., Rockman, M. V., Zimmer, M., Jang, H., Macosko, E. Z., Kruglyak, L., and Bargmann, C. I. (2009). Quantitative mapping of a digenic behavioral trait implicates globin variation in *C. elegans* sensory behaviors. *Neuron* **61**(5), 692–699.

McVeigh, P., Leech, S., Mair, G. R., Marks, N. J., Geary, T. G., and Maule, A. G. (2005). Analysis of FMRFamide-like peptide (FLP) diversity in phylum Nematoda. *Int. J. Parasitol.* **35**, 1043–1060.

Morran, L. T., Cappy, B. J., Anderson, J. L., and Phillips, P. C. (2009). Sexual partners for the stressed: Facultative outcrossing in the self-fertilizing nematode *Caenorhabditis elegans*. *Evolution* **63**(6), 1473–1482.

Nystul, T. G., and Roth, M. B. (2004). Carbon monoxide-induced suspended animation protects against hypoxic damage in *Caenorhabditis elegans*. *Proc. Natl. Acad. Sci. USA* **101**(24), 9133–9136.

Palopoli, M. F., Rockman, M. V., TinMaung, A., Ramsay, C., Curwen, S., Aduna, A., Laurita, J., and Kruglyak, L. (2008). Molecular basis of the copulatory plug polymorphism in *Caenorhabditis elegans*. *Nature* **454**(7207), 1019–1022.

Persson, A., Gross, E., Laurent, P., Busch, K. E., Bretes, H., and de Bono, M. (2009). Natural variation in a neural globin tunes oxygen sensing in wild *Caenorhabditis elegans*. *Nature* **458** (7241), 1030–1033.

Rai, S., and Rankin, C. H. (2007). Critical and sensitive periods for reversing the effects of mechanosensory deprivation on behavior, nervous system, and development in *Caenorhabditis elegans*. *Dev. Neurobiol.* **67**(11), 1443–1456.

Rajpara, S. M., Garcia, P. D., Roberts, R., Eliassen, J. C., Owens, D. F., Maltby, D., Myers, R. M., and Mayeri, E. (1992). Identification and molecular cloning of a neuropeptide Y homolog that produces prolonged inhibition in *Aplysia* neurons. *Neuron* **9**(3), 505–513.

Reddy, K. C., Andersen, E. C., Kruglyak, L., and Kim, D. H. (2009). A polymorphism in *npr-1* is a behavioral determinant of pathogen susceptibility in *C. elegans*. *Science* **323**(5912), 382–384.

Ren, P., Lim, C. S., Johnsen, R., Albert, P. S., Pilgrim, D., and Riddle, D. L. (1996). Control of *C. elegans* larval development by neuronal expression of a TGF-beta homolog. *Science* **274**(5291), 1389–1391.

Roberts, T. M., and Thorson, R. E. (1977). Chemical attraction between adults of *Nippostrongylus brasiliensis*: Description of the phenomenon and effects of host immunity. *J. Parasitol.* **63**(2), 357–363.

Robinson, G. E., Fernald, R. D., and Clayton, D. F. (2008). Genes and social behavior. *Science* **322** (5903), 896–900.

Rogers, C., Reale, V., Kim, K., Chatwin, H., Li, C., Evans, P., and de Bono, M. (2003). Inhibition of *Caenorhabditis elegans* social feeding by FMRFamide-related peptide activation of NPR-1. *Nat. Neurosci.* **6**(11), 1178–1185.

Rogers, C., Persson, A., Cheung, B., and de Bono, M. (2006). Behavioral motifs and neural pathways coordinating O_2 responses and aggregation in *C. elegans*. *Curr. Biol.* **16**(7), 649–659.

Rose, J. K., Sangha, S., Rai, S., Norman, K. R., and Rankin, C. H. (2005). Decreased sensory stimulation reduces behavioral responding, retards development, and alters neuronal connectivity in *Caenorhabditis elegans*. *J. Neurosci.* **25**(31), 7159–7168.

Sambongi, Y., Nagae, T., Liu, Y., Yoshimizu, T., Takeda, K., Wada, Y., and Futai, M. (1999). Sensing of cadmium and copper ions by externally exposed ADL, ASE, and ASH neurons elicits avoidance response in *Caenorhabditis elegans*. *Neuroreport* **10**(4), 753–757.

Schackwitz, W. S., Inoue, T., and Thomas, J. H. (1996). Chemosensory neurons function in parallel to mediate a pheromone response in *C. elegans*. *Neuron* **17**(4), 719–728.

Shen, C., and Powell-Coffman, J. A. (2003). Genetic analysis of hypoxia signaling and response in *C. elegans*. *Ann. N. Y. Acad. Sci.* **995,** 191–199.

Simon, J. M., and Sternberg, P. W. (2002). Evidence of a mate-finding cue in the hermaphrodite nematode *Caenorhabditis elegans*. *Proc. Natl. Acad. Sci. USA* **99**(3), 1598–1603.

Srinivasan, J., Kaplan, F., Ajredini, R., Zachariah, C., Alborn, H. T., Teal, P. E., Malik, R. U., Edison, A. S., Sternberg, P. W., and Schroeder, F. C. (2008). A blend of small molecules regulates both mating and development in *Caenorhabditis elegans*. *Nature* **454**(7208), 1115–1118.

Styer, K. L., Singh, V., Macosko, E., Steele, S. E., Bargmann, C. I., and Aballay, A. (2008). Innate immunity in *Caenorhabditis elegans* is regulated by neurons expressing NPR-1/GPCR. *Science* **322**(5900), 460–464.

Sulston, J. E., and Horvitz, H. R. (1977). Post-embryonic cell lineages of the nematode, *Caenorhabditis elegans*. *Dev. Biol.* **56**(1), 110–156.

Sulston, J. E., Schierenberg, E., White, J. G., and Thomson, J. N. (1983). The embryonic cell lineage of the nematode *Caenorhabditis elegans*. *Dev. Biol.* **100**(1), 64–119.

Tensen, C. P., Cox, K. J., Burke, J. F., Leurs, R., van der Schors, R. C., Geraerts, W. P., Vreugdenhil, E., and Heerikhuizen, H. (1998). Molecular cloning and characterization of an invertebrate homologue of a neuropeptide Y receptor. *Eur. J. NeuroSci.* **10**(11), 3409–3416.

Thomas, J. H., Birnby, D. A., and Vowels, J. J. (1993). Evidence for parallel processing of sensory information controlling dauer formation in *Caenorhabditis elegans. Genetics* **134**(4), 1105–1117.

Troemel, E. R., Kimmel, B. E., and Bargmann, C. I. (1997). Reprogramming chemotaxis responses: Sensory neurons define olfactory preferences in *C. elegans. Cell* **91**(2), 161–169.

White, J. G., Southgate, E., Thomson, J. N., and Brenner, S. (1986). The structure of the nervous system of *Caenorhabditis elegans. Philos. Trans. R. Soc. Lond. B Biol. Sci.* **314**(1165), 1–340.

2 Molecular Social Interactions: *Drosophila melanogaster* Seminal Fluid Proteins as a Case Study

Laura K. Sirot,[1] Brooke A. LaFlamme,[1] Jessica L. Sitnik, C. Dustin Rubinstein, Frank W. Avila, Clement Y. Chow, and Mariana F. Wolfner

Department of Molecular Biology and Genetics, 421 Biotechnology Building, Cornell University, Ithaca, New York 14853, USA

[1]These authors contributed equally

ABSTRACT

Studies of social behavior generally focus on interactions between two or more individual animals. However, these interactions are not simply between whole animals, but also occur between molecules that were produced by the interacting individuals. Such "molecular social interactions" can both influence and be influenced by the organismal-level social interactions. We illustrate this by

Advances in Genetics, Vol. 68

0065-2660/09 $35.00
DOI: 10.1016/S0065-2660(09)68002-0

reviewing the roles played by seminal fluid proteins (Sfps) in molecular social interactions between males and females of the fruit fly *Drosophila melanogaster*. Sfps, which are produced by males and transferred to females during mating, are involved in inherently social interactions with female-derived molecules, and they influence social interactions between males and females and between a female's past and potential future mates. Here, we explore four examples of molecular social interactions involving *D. melanogaster* Sfps: processes that influence mating, sperm storage, ovulation, and ejaculate transfer. We consider the molecular and organismal players involved in each interaction and the consequences of their interplay for the reproductive success of both sexes. We conclude with a discussion of the ways in which Sfps can both shape and be shaped by (in an evolutionary sense) the molecular social interactions in which they are involved.　© 2009, Elsevier Inc.

While studies of social behavior generally focus on observable interactions between individuals, additional "hidden" social interactions occur on the molecular level. These molecular interactions can be considered social in two ways. First, observable social interactions are influenced by molecular interactions (Ellison and Gray, 2009). Second, molecules from different individuals can interact in what we call here "molecular social interactions." The molecular biology of social behavior has thus far been focused primarily on the former: molecular interactions within an animal that either induce or result from social interactions. This approach has successfully identified molecular interactors in rodent and avian affiliative behavior (e.g., reviewed in Adkins-Regan, 2009; Keverne and Curley, 2004), nematode feeding behavior (e.g., reviewed in de Bono and Maricq, 2005), eusocial behavior (e.g., Smith *et al.*, 2008), and Drosophila courtship (e.g., reviewed in Dickson, 2008; Villella and Hall, 2008). However, a complete molecular understanding of social behavior necessitates an understanding not just of how molecules interact within a social animal, but also how "social molecules" interact among animals. Here, we present a case study of such "molecular social interactions" that involves *Drosophila melanogaster* seminal fluid proteins (Sfps) that are produced in the male reproductive tract, and transferred to the female along with sperm during mating. In the case of *D. melanogaster* Sfps, the molecular social interactions are extensive, as gene products in seminal fluid induce short- and long-term changes in females' behavior, physiology, and gene expression, and these changes require interactions of Sfps with female-derived molecules and physiology (e.g., muscle, circulatory, and neural systems). Thus, the male- and female-derived molecules are involved in an inherently social interaction—that is an interaction between two individuals of the same species. Molecular social interactions affect the outcome of individual matings and occur directly between males and between males and females, and indirectly between multiple males that have mated with a given female. As we will discuss, molecular social interactions both shape, and are shaped by, observable behavioral interactions between conspecifics to affect lifetime reproductive success.

Following mating, female *D. melanogaster* display a number of behavioral and physiological changes that impact both male and female reproductive success. For example, after mating, females increase their rates of oogenesis, ovulation, egg-laying, and food intake (e.g., reviewed in Chapman, 2001; Chapman and Davies, 2004; Wolfner, 2009). Sperm from the male are stored in specialized sperm storage organs (Fig. 2.1), and this process may be facilitated by changes in uterine shape beginning at the onset of mating (Adams and Wolfner, 2007; Avila and Wolfner, 2009). For several days, mated females are less likely to accept suitors, actively fleeing or kicking any persistent male (Ringo, 1996; Spieth and Ringo, 1983). Within hours after mating, the female increases expression of several known antimicrobial peptide genes (Kapelnikov *et al.*, 2008b; Lawniczak and Begun, 2004; Mack *et al.*, 2006; McGraw *et al.*, 2004; Peng *et al.*, 2005b), yet the realized immune response that protects the female from infections is reduced (Fedorka *et al.*, 2007). The lifespan of *Drosophila* females is also reduced by mating (Barnes *et al.*, 2008; Chapman *et al.*, 1995; Civetta and Clark, 2000b; Fowler and Partridge, 1989; Wigby and Chapman, 2005).

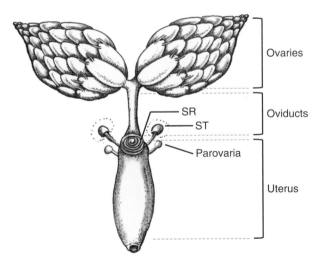

Figure 2.1. Female *Drosophila melanogaster* reproductive tract. During mating, the ejaculate is transferred into the uterus. From here, different components of the ejaculate move to different locations within the female reproductive tract. Some seminal fluid proteins also move out of the reproductive tract and into circulation. SR, seminal receptacle; ST, spermatheca. Drawing by J. Sitnik; for clarity, some parts have been simplified in the figure.

These changes in behavior, physiology, and gene expression may be brought about by the behavioral act of mating, by the transfer of sperm, or by other contents of the seminal fluid. Since males that do not produce sperm still elicit postmating responses in their partners (albeit, weaker and/or more short term; Kalb *et al.*, 1993; Manning, 1962, 1967; Xue and Noll, 2000), nonsperm components of the seminal fluid must be involved in the induction of these responses. In fact, males that transfer sperm but do not transfer Sfps produced in their accessory glands (Fig. 2.1) fail to elicit most postmating responses in females (Kalb *et al.*, 1993; Xue and Noll, 2000). It is known that the ejaculatory duct and ejaculatory bulb also produce secreted proteins that constitute part of seminal fluid, and that some of these proteins are necessary for postmating responses (e.g., Gilbert *et al.*, 1981; Iida and Cavener, 2004; Ludwig *et al.*, 1991; Lung and Wolfner, 2001; Lung *et al.*, 2001; Meikle *et al.*, 1990; Samakovlis *et al.*, 1991; Bretman *et al.*, in press). These results together demonstrate that sperm and Sfps are both required to induce long-term postmating responses in females (Heifetz *et al.*, 2001; Kalb *et al.*, 1993; Kubli, 2003; Manning, 1962, 1967).

Sfps comprise an elaborate intraspecific signaling system. Of the more than 180 predicted extracellular proteins present in the reproductive secretory glands of male *D. melanogaster*, over 100 have been confirmed to be transferred to the female along with sperm (e.g., reviewed in Ravi Ram and Wolfner, 2007a; Chapman, 2008; see also Chintapalli *et al.*, 2007; Findlay *et al.*, 2008, 2009; Takemori and Yamamoto, 2009; Walker *et al.*, 2006). Many of the transferred proteins fall into conserved protein classes found in the seminal fluid of most animals studied to date and include proteases, protease inhibitors, acid lipases, cysteine-rich secretory proteins (CRISPs), and lectins (Mueller *et al.*, 2004; Ravi Ram and Wolfner, 2007a). Other, less-expected, classes of Sfps such as odorant-binding proteins suggest a possible role for small molecules in inducing female postmating responses (Findlay *et al.*, 2008). Odorant-binding proteins are known to shuttle pheromones or other small molecules to odorant receptors in the olfactory system (e.g., reviewed in Pelosi *et al.*, 2005). Presence of predicted odorant-binding proteins in the seminal fluids suggests that they may play a similar shuttling role for molecules once they are within the female reproductive tract. The wide variety of protein classes present in the seminal fluid suggests that Sfps take part in a complex series of interactions within the mated female and do not just fulfill a single simple role.

Upon transfer to females, Sfps target to specific tissues which are likely to relate to their function within the mated female (e.g., Bertram *et al.*, 1996; Heifetz *et al.*, 2000; Lung and Wolfner, 1999; Meikle *et al.*, 1990; Peng *et al.*, 2005a; Ravi Ram *et al.*, 2005; Fig. 2.1). For example, proteins associated with sperm storage and retention have been detected in the female sperm storage organs, and ovulin, which stimulates ovulation, targets to the base of the ovaries (Heifetz *et al.*, 2000; Ravi Ram *et al.*, 2005). Several Sfps, including ovulin, have also been detected in the circulatory system of mated females from where they can gain access to the brain and/or endocrine systems (Lung and Wolfner, 1999; Meikle *et al.*, 1990; Pilpel *et al.*, 2008; Ravi Ram *et al.*, 2005) and thus, potentially, affect female behavior. Further studies of the targets of Sfps may help to uncover their functions in the mated female.

D. *melanogaster* Sfps provide an excellent model system in which to investigate molecular social interactions, due to the powerful tools available in this species. Mutant or transgenic males in which Sfps are increased, decreased, or eliminated can be used to dissect the effect(s) of particular Sfps on female postmating responses (e.g., Bretman *et al.*, in press; Chapman *et al.*, 2003; Gilbert *et al.*, 1981; Herndon and Wolfner, 1995; Iida and Cavener, 2004; Liu and Kubli, 2003; Mueller *et al.*, 2008; Neubaum and Wolfner, 1999; Ravi Ram and Wolfner, 2007b; Ravi Ram *et al.*, 2006; Wong *et al.*, 2008a). A large collection of freely available genomic databases (e.g., FlyBase; FlyAtlas, Chintapalli *et al.*, 2007) facilitate rapid progress as well. These techniques and tools, along with studies associating allelic variation in Sfps with variation in their effects, have led to a greater understanding of the molecular social interactions taking place between all of the players involved in *Drosophila* mating (e.g., reviewed in Wolfner, 2009). Furthermore, studies of D. *melanogaster* Sfps are likely to provide insights into the molecular social interactions of other species given that Sfps impact female postmating responses across a wide taxonomic range (e.g., reviewed in Gillott, 2003; Poiani, 2006).

We will use two particularly well-studied Sfps, the sex peptide (SP) and ovulin, as examples in the following sections to illustrate the way in which Sfps act as molecular mediators for social interactions. SP is a small peptide that affects female response to male courtship, oogenesis, and her ovulation, immune response, feeding, and juvenile hormone production (Carvalho *et al.*, 2006; Chapman *et al.*, 2003; Domanitskaya *et al.*, 2007; Kubli, 2003; Liu and Kubli, 2003; Moshitzky *et al.*, 1996). Ovulin is a large prohormone that increases ovulation during the first 24 h after mating. Further details of both these proteins, as well as the social context in which they exert their functions, are discussed herein.

While over 180 known or putative D. *melanogaster* Sfps have been identified, only one female receptor to an Sfp is known: the SP receptor (SPR), a G-protein-coupled receptor expressed in the female reproductive tract and nervous system (Yapici *et al.*, 2008). However, we expect that many Sfps interact with female-derived proteins. Some female-derived proteins that play a role in female postmating behavior and physiology have been identified and will be discussed in this review, but their interactions with Sfps remain speculative at this time.

Several approaches have been used to identify genes in females whose products mediate response to, are regulated by, or otherwise interact with, Sfps. Proteins produced in the female sperm storage organs have been identified and have the potential to interact with Sfps (Allen and Spradling, 2008; Lawniczak and Begun, 2007; Prokupek *et al.*, 2008, 2009). Microarray data from whole flies, heads, or reproductive tract tissues have shown that different aspects of mating, including Sfps, cause a transcriptional response in the female after mating (Innocenti and Morrow, 2009; Kapelnikov *et al.*, 2008b; Lawniczak and Begun, 2004, 2007; Mack *et al.*, 2006; McGraw *et al.*, 2004, 2008, 2009; Peng *et al.*, 2005b), though it is not likely that most initial postmating responses are due to mating-induced transcription. Transcriptional changes of the largest magnitude are seen by about 6–8 h after mating, a time by which most Sfps are no longer detectable in the female. Therefore, Sfps may set into motion the transcriptional modification of the female, but the genes regulated by these modifications are less likely to encode Sfp-interacting proteins than the genes expressed by the female prior to mating. Nevertheless, these mating-regulated genes likely are players in the next steps of the molecular social interactions. To fully understand the molecular social interactions in which Sfps are involved, we must identify female interactors, their functions, and how they have coevolved with their male-derived partners.

I. *DROSOPHILA* SFPS AND MOLECULAR SOCIAL INTERACTIONS: AN INTERPLAY IN FOUR ACTS

In this section, we present four examples of molecular social interactions in the format of acts in a play involving *D. melanogaster* Sfps and those molecules with which they interact, derived from different actors (a female and her past, present, and potential future mates), that occur at different times (before, during, and after copulation), in different settings (outside and inside the female's body, within the female reproductive tract, nervous and circulatory systems), and on different time scales (immediately and over the course of evolution). We also discuss the potential influences of each of these interactions on male and female reproductive success. For many of the processes that we describe, there are other known molecular actors, but they are not known to interact with Sfps and therefore are not included in this review for simplicity. The curtain is rising.

A. Setting the stage: Transfer and fate of sperm and Sfps

Copulation duration: Copulation duration is generally ~ 20 min in *D. melanogaster* (Gilchrist and Partridge, 2000) and is influenced by several factors that include: female cuticular hydrocarbons (Marcillac and Ferveur, 2004), female mating history (Friberg, 2006; Singh and Singh, 2004), male and female body size (LeFranc and Bundgaard, 2000), and the social environment before and during copulation (see below; Bretman *et al.*, 2009; Wigby *et al.*, 2009).

Copulation duration in *D. melanogaster* is generally longer than necessary for sperm transfer and may be subject to sexual conflict (Gilchrist and Partridge, 2000; Mazzi *et al.*, 2009).

Mating plug: Shortly after the start of mating, Sfps from the male's ejaculatory bulb begin to form a mating plug in the female's uterus (Ludwig *et al.*, 1991; Bretman *et al.*, in press). By the end of mating, Sfps from the male accessory glands will have completed the mating plug (Lung and Wolfner, 2001). The purpose of the mating plug is unknown: it is unlikely that it physically and permanently blocks subsequent inseminations of the female as it is only present for a few hours, yet recent evidence suggests that an Sfp in the mating plug may prevent females from remating shortly after their initial mating (Bretman *et al.*, in press). The mating plug may also be a way to assist with sperm storage, including potentially to protect the male's ejaculate from being expelled by the female before sperm storage is completed.

Sperm: The time of sperm transfer varies between individual matings and between different fly strains (Fowler, 1973), but appears to be complete by 8 min after the start of mating (Fowler, 1973; Gilchrist and Partridge, 2000; Lung and Wolfner, 2001). Sperm transfer itself takes only about 1 min to complete (Gilchrist and Partridge, 2000). Upon transfer to the female, sperm move through the uterus and into specialized sperm storage organs, the paired sper-mathecae and the seminal receptacle. The sperm storage process begins during mating and continues for a few hours (Bloch Qazi *et al.*, 2003). Sperm can remain viable in the female's sperm storage organs for approximately 2 weeks (Kaufmann and Demerec, 1942; Lefevre and Jonsson, 1962). Sperm from multiple males can co-occur in the sperm storage organs of females because *D. melanogaster* females, as in other animal species, commonly mate with multiple males in nature (e.g., see Birkhead and Moller, 1998; Boorman and Parker, 1976; Harshman and Clark, 1998; Imhof *et al.*, 1998; Simmons, 2001). When two males mate succes-sively with the same female, there is an opportunity for sperm competition as the gametes of the different males compete to fertilize the female's eggs.

B. Act 1: To mate or not to mate

Setting: The mating arena
Time: Before copulation
Organismal actors: Male, female, female's previous mate (in absentia)
Known molecular actors: SP and SPR, four supporting Sfps (cuticular and ejacula-tory bulb pheromones which also participate in this Act will not be discussed here; for review see, e.g., Dickson, 2008; Ferveur, 1997; Greenspan and Ferveur, 2000; Hall, 1994; Markow and O'Grady, 2005; Villella and Hall, 2008).

Effect: Sfps received from one mate affect the outcome of subsequent intersexual interactions

When a male and female *D. melanogaster* encounter each other, the outcome is often a series of courtship steps (Greenspan and Ferveur, 2000; Hall, 1994) followed by copulation. But, what factors determine whether a male will court and attempt to copulate and how a female responds to the solicitations by males? In *D. melanogaster*, the decision by males to court and the response of females to male courtship depends, in part, on female mating status: males that have previously courted mated females are less likely subsequently to court a mated female than an unmated female, and mated females are less likely than unmated females to copulate with a courting male. The difference in male courtship behavior in response to female mating status is due, in part, to molecular social interactions between the male, the female, and the female's previous mates (e.g., Dickson, 2008; Ejima *et al.*, 2007; Ferveur, 1997; Greenspan and Ferveur, 2000; Markow and O'Grady, 2005; Scott, 1986; Tompkins *et al.*, 1983; Villella and Hall, 2008; Yew *et al.*, 2009), but male courtship intensity does not depend on whether a female received Sfps from her previous mate (Tram and Wolfner, 1998).

Mating-dependent changes in female responses to male courtship (i.e., whether the female copulates or not) have also been attributed to molecular social interactions. For example, females that received the Sfp SP during mating were much more likely to exhibit rejection behaviors (e.g., extrude ovipositor; Fig. 2.2) and less likely to copulate with a courting male than females that did

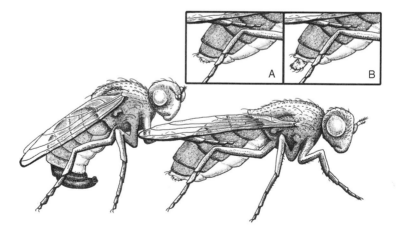

Figure 2.2. Mating attempt by male and (insets) position of female abdomen when she is receptive (A) and unreceptive (B; ovipositor extruded) to the mating attempt. Drawing by J. Sitnik.

not receive SP (Aigaki *et al.*, 1991; Chapman *et al.*, 2003; Chen *et al.*, 1988; Liu and Kubli, 2003; Soller *et al.*, 2006; Yapici *et al.*, 2008). Recently, another protein, PebII, has been found to reduce female receptivity immediately after mating (Bretman *et al.*, submitted for publication). SP interacts with a G-protein-coupled receptor (SPR) in the female nervous system to effect female response to male courtship (Yapici *et al.*, 2008). SP and SPR act through approximately 6–8 *fruitless-* and *pickpocket*-expressing multidentritic sensory neurons (Hasemeyer *et al.*, 2009; Yang *et al.*, 2009) located along the reproductive tract. Without SPR, females do not show the SP-induced postmating changes in egg-laying or in rejection behavior, indicating the primary role for this molecule in these aspects of the female's postmating responses (Yapici *et al.*, 2008).

The effects of SP are long-lasting because SP is maintained in the mated female's sperm storage organs for days after mating and is gradually released (Peng *et al.*, 2005a). SP binds to sperm, which apparently stabilizes the peptide. As long as the female contains sperm with bound SP, she is likely to display rejection behavior in response to mating attempts by other males. This long-term effect of SP allows a male to affect his mate's social interactions, in absentia, for several days after mating has ended (Chapman *et al.*, 2003; Liu and Kubli, 2003). At least four additional Sfps are necessary for this facet of SP function. These Sfps, the predicted CRISP CG17575, the predicted protease CG9997 and the gene duplicate lectins CG1652 and CG1656 work in an interdependent network to localize SP to the sperm storage organs of the females and/or bound to sperm (Ravi Ram and Wolfner, 2009). In contrast to SP, these four Sfps do not persist in females for days. Rather, they exert their effects soon after mating and then disappear. It is the persistence of SP that maintains the female's postmating responses of egg-laying and remating rejection. If SP, or any of the network Sfps, is not transferred to a female, she resumes being willing to mate with a subsequent courting male within a day of her first mating (Ravi Ram and Wolfner, 2007b).

Whether, when, and how frequently a female *D. melanogaster* remates affects not only the reproductive success of her previous and potential subsequent mates but also her own reproductive success. On the male side, it is generally in a female's previous mate(s)' interest that she either delays or forgoes remating (so that his sperm will not be displaced), while it is in her current suitor's interest that she does remate (so that he can sire offspring). On the female side, as in some other species, repeated mating can be costly in terms of lifetime reproductive success (Arnqvist and Nilsson, 2000; Chapman *et al.*, 1995, 2000; Fowler and Partridge, 1989), but repeated matings can also increase the reproductive success of a female's daughters (Priest *et al.*, 2008a) or benefit the female in other ways (e.g., avoiding infertile matings or genetic incompatibility; Arnqvist and Nilsson, 2000). Because matings can be costly to females (Fowler and Partridge, 1989), there is the potential for intersexual conflict between females and their suitors over remating (Arnqvist and Rowe, 2005).

The example described in Act 1 demonstrates that the transfer of molecules during mating serves as a means of interaction between a female and her mate, between a female and the subsequent males she encounters, and between the female's multiple mating partners. Thus, the outcome of an encounter between male and female can be said to be influenced by both inter- and intrasexual molecular social interactions and can potentially result in both inter- and intrasexual conflict over female remating.

C. Act 2: Should sperm stay or should they go?

Setting: The female reproductive tract
Time: During and after copulation
Organismal/cellular actors: Female, sperm
Known molecular actors: Acp36DE and other Sfps, proteins from the female sperm storage organs
Effect: Sfps affect conformation of uterus, sperm storage, and success in sperm competition

Once copulation begins, a series of molecular social interactions unfolds within the female's reproductive tract. A subset of these interactions determines the fate of sperm in the female reproductive tract and, thereby, influences male reproductive success in two ways: (i) storage and subsequent release of sperm from storage determine which male's sperm fertilize a female's eggs and (ii) sperm storage is necessary for the persistence of postmating changes in female behavior (e.g., by serving as a source of SP; see Act 1 above; Manning, 1962, 1967; Peng et al., 2005a; see Bloch Qazi et al., 2003, for review). Thus, getting sperm into storage is of vital importance to males, and male contributions appear to play an active role in making sure this happens. Time is also of the essence: sperm that remain unstored in the female's uterus when she lays her first egg postmating may be expelled from the female (Fowler, 1973; Gilbert et al., 1981), creating a potential tradeoff between the timing of sperm storage and egg output.

The mechanism by which sperm move through the female reproductive tract presents some unusual considerations. Unlike mammalian sperm, *D. melanogaster* sperm are extremely long relative to the female's body size (roughly half the length of the entire female; Pitnick et al., 1995) and thus may be constrained from propelling through the female reproductive tract. Molecular social interactions appear to play a role in solving this conundrum. Recent evidence suggests that sperm movement through the female reproductive tract is mediated by uterine conformational changes that themselves are mediated by Sfps (Adams and Wolfner, 2007; Avila and Wolfner, 2009). Before mating, the female reproductive tract is in a "closed" conformation: the lumen is tightly compacted and folded into an S-shape and the entrance to the sperm storage organs is physically blocked by a flap of the uterine wall (Adams and Wolfner, 2007).

Within the first few minutes of mating, the posterior uterus begins to open up and eventually the entrance to the sperm storage organs becomes unblocked, with these changes continuing for approximately 45 min after the start of mating until the entire uterine lumen is open. During this time, the sperm mass is moved up the reproductive tract (perhaps as a result of the conformational changes) and the sperm mass becomes situated at the site from which sperm can enter storage.

The changes in uterine conformation are not simply the result of the mechanics of mating, nor do they require sperm to take place (Adams and Wolfner, 2007). Sfps, and not sperm, are required for these conformational changes to occur (Adams and Wolfner, 2007). One particular Sfp, the glycoprotein Acp36DE, is needed for the reproductive tract to proceed past the mid-stages (Avila and Wolfner, 2009). In the absence of Acp36DE, the female tract only opens part way, retaining a constriction near the anterior end of the uterus, which appears to prevent the majority of sperm from being actively stored. Consistent with this hypothesis, Acp36DE is also required for efficient and complete sperm storage, with only 10–50% of the sperm stored in a wild-type mating being stored when females mate to Acp36DE null males (Bloch Qazi and Wolfner, 2003; Neubaum and Wolfner, 1999). Presumably as a result of its effects on sperm storage, Acp36DE also impacts a male's ability to compete for fertilizations (Chapman et al., 2000).

The female molecules and physiology involved in the uterine conformational changes are currently unknown. One plausible mechanism by which Sfps effect uterine conformational changes is by stimulating the female's nervous system to control the release of neuromodulators that induce muscle contraction or relaxation. The uterus is surrounded by both circular and longitudinal muscles (Adams and Wolfner, 2007; Heifetz and Wolfner, 2004; Middleton et al., 2006). Mating enhances muscle differentiation and innervation in the female reproductive tract and causes vesicle release from nerve termini that innervate the female reproductive tract (Heifetz and Wolfner, 2004; Kapelnikov et al., 2008a). Sfps, especially Acp36DE, should be tested for a role in inducing such vesicle release.

Uterine conformational changes are not the only way in which male and female molecules may impact sperm storage. For example, glucose dehydrogenase, a protein produced in the reproductive tracts of both males and females, affects both the number and distribution of sperm stored in the paired spermathecae (Iida and Cavener, 2004). Another Sfp, Esterase-6, influences the timing and rate of sperm storage (Gilbert et al., 1981). Sperm proteins, such as a polycystin-2-like protein found on the flagella, are also necessary for the storage of sperm that have been transferred to the female (Gao et al., 2003; Watnick et al., 2003).

The influence of molecular social interactions on sperm use patterns does not end with sperm entry into storage. Sperm must be maintained in storage and then released at an appropriate time to fertilize eggs that are released from the ovary into the oviducts and uterus (see Act 3). Females must receive Sfps to use sperm from a male. Females that do not receive Sfps during mating will store a small number of sperm, but those sperm will not be used to fertilize eggs (Hihara, 1981; Xue and Noll, 2000). Interestingly, Sfps received by a female in a previous or subsequent mating can partially restore the fertilizing ability of another mate's sperm (Xue and Noll, 2000). In *in vitro* assays, D. *melanogaster* seminal fluid also helps to maintain sperm viability over a short-time scale (1 h, Holman, 2009), but it is not known whether this is also the case within the mated female. We do not yet know the male and female molecules involved in maintaining sperm viability in the female, but secretions from the spermathecae appear to be necessary for prolonged sperm viability (Allen and Spradling, 2008; Anderson, 1945; Boulétreau-Merle, 1977). For example, sperm release from storage depends, in part, on spermathecal proteins (Iida and Cavener, 2004) and on the presence of eggs in the female reproductive tract: the transition from sperm storage to sperm release is delayed in egg-less females (Bloch Qazi and Wolfner, 2003). Retention and release of stored sperm also depends, in part, on Sfps. Six Sfps have been identified that are necessary for retention or release of sperm from the female sperm storage organs (Gilbert *et al.*, 1981; Ravi Ram and Wolfner, 2007b; Wong *et al.*, 2008a). Interestingly, four of the Sfps that affect sperm release are the same proteins described in Act 1 that are needed to get SP into the female's sperm storage organs (Ravi Ram and Wolfner, 2009). Mechanistic connections (if any) between the multiple roles of these proteins are not yet known. Furthermore, the interactions between male- and female-derived molecules that influence sperm release have yet to be explored.

The storage, maintenance, and timely release of sperm in the female's sperm storage organs have important consequences for both male and female reproductive success. Clearly, a male will have the opportunity to fertilize a female's eggs only if his sperm are stored, maintained, and released when appropriate. For a female, storing sperm allows her eggs to be fertilized for at least 2 weeks (the duration of sperm storage) even if she does not encounter a male (Kaufmann and Demerec, 1942; Lefevre and Jonsson, 1962). Sperm storage also allows a female to delay egg-laying until she finds suitable environmental conditions for offspring development. Females can also benefit by storing sperm of multiple males, which provides opportunities for both sperm competition (Boorman and Parker, 1976; Simmons, 2001) and cryptic female choice (Eberhard, 1996).

D. Act 3: *Pas de deux*

Setting: The female reproductive tract
Time: During and after copulation
Organismal/cellular actors: Eggs, neurons, muscles
Known molecular actor: Ovulin and Supporting Sfps, Octopamine, OAMB

Effect: Processing of an ovulation-stimulating Sfp requires both male and female contributions

In insects, ovulation behavior is especially sensitive to male–female interactions, as ovulation dramatically increases after mating. Unmated *D. melanogaster* females will lay a few unfertilized eggs per day (~2), whereas a mated female will lay several dozen eggs within the first day after mating (~50). This change in number of eggs laid is due to changes in both oogenesis and ovulation and both of these processes are influenced by Sfps (Chapman, 2001; Prout and Clark, 2000). The stimulation of oogenesis appears to result, in part, from an SP-induced increase in juvenile hormone levels and from oogenesis progressing past an arrest point; these will not be discussed further here (Moshitzky *et al.*, 1996; Soller *et al.*, 1999). The changes in ovulation rate are influenced by a different Sfp, ovulin, and are discussed below.

During ovulation, mature oocytes are released from the ovary and pass through the oviduct via muscle contractions requiring coordination among the nervous system, muscle, and epithelium of the female reproductive tract (Lange, 2009; Middleton *et al.*, 2006; Rodríguez-Valentín *et al.*, 2006). This coordination ultimately results in oviposition on the substrate (Yang *et al.*, 2008). Female molecules necessary for the movement of eggs through the female reproductive tract have been identified. For example, the biogenic amine octopamine and one of its receptors (OAMB) are required for ovulation (Cole *et al.*, 2005; Lee *et al.*, 2003, 2009; Monastirioti, 1999, 2003). Neurons producing neuromodulators (e.g., octopamine) and neuropeptides (e.g., ILP7) innervate the muscles and epithelium of the female reproductive tract (Lee *et al.*, 2009; Middleton *et al.*, 2006; Monastirioti, 2003; Rodríguez-Valentín *et al.*, 2006; Yang *et al.*, 2009). Sfps modulate release of vesicles from nerve termini in the female reproductive tract (Heifetz and Wolfner, 2004), which may include levels of octopamine release. Thus, Sfps may regulate ovulation by modulating the octopamine system. Furthermore, vesicle transport proteins in the p24 family, are necessary for oviposition in females. In female *D. melanogaster* mutant for these genes, failure to oviposit results in eggs remaining lodged in the uterus, with further mature eggs backing up in the upper reproductive tract (Bartoszewski *et al.*, 2004; Boltz *et al.*, 2007; Carney and Taylor, 2003). Whether and how these proteins interact with Sfps is currently unknown.

On the male side, the transfer of sperm and Sfps is required to trigger the full response of increased egg-laying (Heifetz *et al.*, 2001). Ovulin contributes to the dramatic increase in egg-laying rates observed within 24 h after mating (Heifetz *et al.*, 2000; Herndon and Wolfner, 1995). Specifically, as described in the introduction, ovulin increases ovulation rate (Heifetz *et al.*, 2000). Ovulin's mechanism of action is currently an area of active study. After mating, most of the transferred ovulin remains in the reproductive tract, and targets to the base of the ovaries (Heifetz et al., 2000; Monsma et al., 1990). However, a substantial amount (about 10% of the ovulin that the female receives) crosses a special transiently permeable region of the reproductive tract wall to enter the circulatory system (Lung and Wolfner, 1999; Monsma et al., 1990). In principle, ovulin could act from either place to stimulate ovulation. From its site at the base of the ovary it might directly stimulate muscle contraction or the release or action of neuromodulators such as octopamine (Lee et al., 2003, 2009; Monastirioti, 2003); from the circulatory system, it could act from the outside on the reproductive tract musculature to effect changes in contraction or could act indirectly by binding to neural or endocrine targets removed from the reproductive system and causing activity in those that then affected the reproductive tract.

Ovulin is the object of a particularly interesting molecular social interaction that may affect ovulation. This interaction appears to be a *pas de deux* between male and female molecules resulting in the systematic processing of ovulin into smaller pieces (Monsma and Wolfner, 1988; Park and Wolfner, 1995). Upon transfer, ovulin within the female reproductive tract is processed into several cleavage products (as are several other Sfps; Bertram et al., 1996; Ravi Ram et al., 2006; Fig. 2.3). It is unknown what role the processing of ovulin plays in its effect on ovulation, but indirect evidence suggests that, as with prohormones in other species (e.g., Chen and Raikhel, 1996; Garden et al.,

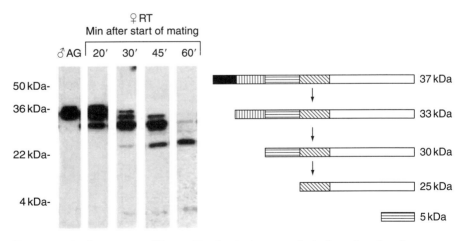

Figure 2.3. Ovulin processing. Western blot showing intact ovulin in the male and ovulin processing products over time after the start of mating (ASM) in the female reproductive tract. Schematic of ovulin processing is on right (schematic adapted from Park and Wolfner (1995); 37 kDa is the size of intact ovulin; in some places on the blot, bands are broad due to glycosylation variants that are not shown in the schematic). (See Page 1 in Color Section at the back of the book.)

1998; Hook *et al.*, 1994), processing may increase ovulin's activity: expression of ovulin's cleavage products in unmated females stimulates a higher rate of ovulation than does expression of full-length ovulin (Heifetz *et al.*, 2005). One of these bioactive cleavage products contains regions with sequence similarities to the egg-laying hormone of the mollusk *Aplysia* (Monsma and Wolfner, 1988). The different ovulin cleavage products appear in the same pattern and over a reproducible time scale in matings between wild-type flies, beginning ~10 min after the start of mating and ending about 2 h later, by which time ovulin is fully processed (Park and Wolfner, 1995). Thus, this time-dependent processing of ovulin could potentially provide a mechanism by which ovulation rate is regulated.

Ovulin cleavage requires male contributions, but only occurs after ovulin has been transferred to the female (Park and Wolfner, 1995), suggesting that female contributions or infrastructure are necessary for processing to occur. One molecule necessary for ovulin's cleavage is the male-derived Sfp CG11864 (Ravi Ram *et al.*, 2006). This protein is a predicted metalloprotease that itself is cleaved from a predicted inactive form to the predicted active form while in transit to the female, by one or more other Sfps (Ravi Ram, LaFlamme, Wolfner, unpublished data). Though CG11864 attains its predicted active form while still inside the male, there is no evidence that it can cleave ovulin before it is inside the female's reproductive tract. As yet unidentified female-derived factors may also be required for ovulin's cleavage. This processing represents a fascinating example of a highly coordinated molecular social interaction (and biochemical pathway) that begins in the male, ends within the female, and may impact both male and female reproductive success. But, our knowledge of ovulin processing also raises more questions. Why is this process so tightly regulated? Are the contributions of males and females to ovulin processing acting in conflict or in cooperation? How does this molecular social interaction affect the reproductive success of both sexes?

Before we can fully understand the molecular social interactions that occur during the proteolytic processing of ovulin, we must identify the female factors that regulate this processing. The most promising molecular candidates for ovulin processing are proteolysis regulators. Several studies of female reproductive tract proteins have discovered female-secreted proteases and protease inhibitors that are regulated in response to mating (Arbeitman *et al.*, 2002; Chintapalli *et al.*, 2007; Kapelnikov *et al.*, 2008b; Lawniczak and Begun, 2004, 2007; Mack *et al.*, 2006; McGraw *et al.*, 2004; Swanson *et al.*, 2004). It is likely that some of these are involved in proteolytic pathways that begin in the male and end in the female, such as that for the processing of ovulin, and these proteins are currently under investigation. Other female-derived influences on ovulin processing could potentially include small molecules and the physiological environment of the female reproductive tract (e.g., pH and other ionic conditions).

Irrespective of the specific molecular actors involved, it is clear that both females and their mates influence the female's rate of ovulation shortly after mating. Thus, it is not surprising that ovulation rate at this time has important consequences for both sexes. At this point, it is important to recall from Act 2 that the first egg that moves through the uterus after mating appears to push out sperm that have not moved into storage. Therefore, ovulation shortly after mating influences not only progeny production but also subsequent male fertilization success. In a female's first mating, the interest of the sexes in relation to ovulation rate may be congruent, as ovulation shortly after mating appears to clear stale eggs from the uterus, stimulate oogenesis, and coordinate egg production and sperm storage (Chapman *et al.*, 2001). However, in subsequent matings, ovulation rate shortly after mating could be the subject of intersexual conflict. Males may benefit by temporarily delaying ovulation to maximize sperm storage, whereas females may benefit by adjusting the timing of ovulation to influence the subsequent fertilization success of her mate (e.g., by ovulating quickly to eject a male's sperm or delaying ovulation to maximize sperm storage). Thus far, ovulation rate has only been examined after a female's first mating. Future research should test for evidence of cryptic female choice (Eberhard, 1996) and sexual conflict (Arnqvist and Rowe, 2005) over ovulation rate in subsequent matings. Certainly, the fact that the proportion of unfertilized eggs laid increases after a female's subsequent mating (Prout and Clark, 2000) suggests that coordination of the release of sperm and eggs is being modulated by the female, the male, or both.

E. Act 4: Two is company, three is a crowd

Setting: The mating arena
Time: During and after copulation
Organismal actors: Entire cast!
Known molecular actors: SP, SPR, ovulin, probably other Sfps and female-derived molecules
Effect: Social environment and female mating status affect mating duration, amount of Sfps transferred, and male competitive reproductive success

Although *D. melanogaster* copulations may look very much the same to the outside observer, copulations can vary substantially in the amount of Sfps transferred (Sirot *et al.*, 2009). The causes and consequences of this variation have only recently begun to be explored in *D. melanogaster*, where the social environment influences the amount of Sfps transferred (Bretman *et al.*, 2009; Friberg, 2006; Linklater *et al.*, 2007; Wigby *et al.*, 2009). In particular, the quantity of Sfps transferred by a male depends on whether other males are present before and during copulation: males transfer more of the two Sfps tested (SP and ovulin; Fig. 2.4) when they are in the presence of another male before

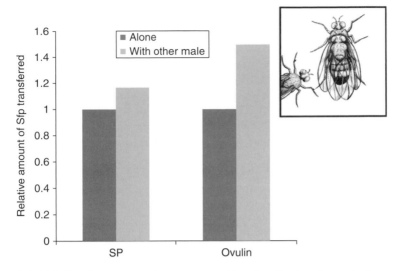

Figure 2.4. Male *D. melanogaster* transfer more sex peptide and ovulin when they are in the presence of another male before and during copulation than when they are alone. Data figure adapted from Wigby *et al.* (2009). Drawing by J. Sitnik; for clarity, some body parts have been simplified in the figure.

and during copulation than when they are alone with the female (Wigby *et al.*, 2009). Thus, males can adjust the amount of Sfps they transfer in response to the potential level of sperm competition. Males also appear to adjust Sfp allocation in response to another metric of sperm competition level: whether the male perceives that the female has previously mated or not (Friberg, 2006). Such "strategic allocation" of Sfps is predicted by theory (Cameron *et al.*, 2007) and has also been demonstrated for the sperm component of the ejaculate in many species (reviewed in Wedell *et al.*, 2002). The influences of other social environmental variables (e.g., operational sex ratio, female and male density) on Sfp transfer have yet to be systematically explored, but some of the variation in Sfp transfer by males results from depletion with successive matings and replenishment thereafter (DiBenedetto *et al.*, 1990; Sirot *et al.*, 2009).

Variation in Sfp transfer appears to result in corresponding variation in female postmating responses. In the extreme situation in which females receive no Sfps derived from the male accessory glands during mating, females show little or no changes in postmating behavior and males fertilize no eggs (Xue and Noll, 2000). In less extreme cases, for example, when males transfer more Sfps in response to potential sperm competition (Bretman *et al.*, 2009; Wigby *et al.*, 2009) or selection for large accessory gland size (Wigby *et al.*, 2009), female postmating responses are more pronounced (e.g., higher egg production, longer delay in remating) and males sire more offspring. Similarly, females that receive lower-than-normal amounts of Sfps because their mates had recently mated show less pronounced postmating responses than females mated to unmated males (Hihara, 1981).

The consequences of variation in Sfp transfer for male reproductive success appear clear: males experience higher competitive reproductive success when they transfer more Sfps (Wigby _et al._, 2009). The consequences of variation in Sfp transfer for females are more complex. Lifetime offspring production of females that repeatedly receive SP is lower than that of females that mate but do not receive SP (Wigby and Chapman, 2005). However, females that receive the full suite of Sfps produce daughters with higher lifetime offspring production than females that mate but do not receive Sfps (Priest _et al._, 2008b; but see also Long _et al._, 2009; Priest _et al._, 2009). More research is needed to determine how subtle variation in amount of Sfps received affects female offspring and grand-offspring production.

Finally, it is important to point out that, since Sfp production appears to be both costly (Wigby _et al._, 2009) and limiting (Hihara, 1981; Sirot _et al._, 2009), male investment in individual ejaculates may be greater than previously appreciated (Cordts and Partridge, 1996). These costs, together with the costs of courtship (Cordts and Partridge, 1996), may help to explain the benefits to males of selectively pursuing (or not pursuing) certain females based on phenotype or mating status (e.g., Byrne and Rice, 2006) and relates to the question posed in Act 1: to mate or not to mate.

II. BEHIND THE SCENES: EVOLUTIONARY DYNAMICS

Behind the scenes of the main mating arena, a number of important factors have been setting the stage. Some of these are developmental genes that determine the sex of the fly, such as _Sxl_ (e.g., reviewed in Christiansen _et al._, 2002; Cline and Meyer, 1996; Manoli _et al._, 2006) or its downstream effectors, such as _dsf_ and _fru_, that are needed for sex-specific behavior and neural development (e.g., reviewed in Dickson, 2008; Manoli _et al._, 2006, Shirangi and McKeown, 2007; Yamamoto, 2007). Other genes are required for proper development of the genital tract and fertility; examples include: _HR39_ (Allen and Spradling, 2008) or _lz_ (Anderson, 1945) in females, and a late function of _prd_ (Xue and Noll, 2002) in males. Though the actions of these genes are thus ultimately important in mating interactions of _Drosophila_, they have a limited role in the types of molecular social interactions we have discussed and will not be discussed further here. On the other hand, evolutionary processes shape all genes in the genome and have a specific role in affecting mating behavior and related social interactions in _Drosophila_. These are discussed further below.

Molecular social interactions, such as those discussed earlier, influence not only the individual interacting flies but also the evolution of the interacting molecules. Over time, evolutionary responses to the interactions between Sfps and female molecules may leave signatures on the genes' sequences themselves (some examples in *Drosophila* include: Begun and Lindfors, 2005; Clark, 2002; Clark *et al.*, 2006; Haerty *et al.*, 2007; Kelleher *et al.*, 2007; Mueller *et al.*, 2005; Schully and Hellberg, 2006; Wong *et al.*, 2008b; see Panhuis *et al.*, 2006; Swanson and Vacquier, 2002; for reviews). Genes encoding *Drosophila* Sfps, like reproduction-related genes across a wide range of taxa, are more likely to show evidence of positive selection than are most other groups of genes, with the notable exception of immunity-related genes (Begun and Lindfors, 2005; Clark, 2002; Dean *et al.*, 2008; Haerty *et al.*, 2007; Kelleher *et al.*, 2007; Mueller *et al.*, 2005; Schully and Hellberg, 2006; Swanson *et al.*, 2001; Wong *et al.*, 2008b; See Clark *et al.*, 2006; Panhuis *et al.*, 2006; Swanson and Vacquier, 2002; for review). It has been suggested that rapid evolution of these genes may result from sexual selection (e.g., cryptic female choice, sperm competition) and/or sexual conflict (reviewed in Swanson and Vacquier, 2002).

An example of the interplay between mechanisms of action and evolutionary dynamics is ovulin. Earlier, we presented ovulin processing as an example of a highly coordinated process between the sexes that begins in the male and ends in the female. Thus, ovulin's action to increase ovulation soon after mating would appear to benefit both sexes, but further consideration raises some questions. Is ovulin a mechanism by which the male controls the female's physiology, or does the female take advantage of the male's ovulin contribution for her own benefit? Has ovulin's evolution been driven by sexual conflict? Or, is ovulin's function truly beneficial to both sexes? Though we currently cannot answer many of the important questions about ovulin, such as why it is processed or how processing of this protein affects its function, we may be able to gain some clues from the evolutionary history of the protein.

Consistent with its important function in reproduction, ovulin contains short regions that are highly conserved between *Drosophila* species (Wong *et al.*, 2006, submitted for publication), with its C-terminus in particular containing several conserved motifs. Genetic and biochemical studies indicate that these motifs mediate self-association of ovulin molecules, evidenced by observations of ovulin multimers *in vivo* in *D. melanogaster*, as well as *in vitro* between ovulin molecules from different species that contain these motifs (Wong *et al.*, 2006, submitted for publication). These self-association domains are within a portion of ovulin that has ovulation-inducing activity, although it is as yet unknown whether self-association is needed for ovulin's function.

Interspersed around the highly conserved backbone of ovulin are rapidly evolving regions. In fact, ovulin is one of the most rapidly evolving proteins in the *Drosophila* genome (Aguadé, 1998; Aguadé *et al.*, 1992; Tsaur and Wu, 1997;

Tsaur *et al.*, 1998). Amino acid divergence between ovulin from *D. melanogaster* and its close relative, *Drosophila simulans*, is about 15%, compared to a 1–2% average for all other genes (Andolfatto, 2005; Kern *et al.*, 2004; Tamura *et al.*, 2004). There is evidence that some of ovulin's amino acid divergence results from positive selection, but the selective forces have not been identified (Aguadé, 1998; Aguadé *et al.*, 1992; Fay and Wu, 2000; Tsaur and Wu, 1997; Tsaur *et al.*, 1998). The ovulin-type pattern of highly divergent protein regions on an otherwise conserved backbone may be the result of sexual conflict, resulting from sexual selection, acting on proteins required for essential biological processes, such as those discussed in Acts 2 and 3.

Given the importance of Sfp genes to reproductive success, positive selection should drive the most advantageous alleles to fixation. We would, therefore, expect any variation in genes under positive selection to be transient. However, in contrast to this prediction, nucleotide variation is maintained in some Sfps, like ovulin (Aguadé *et al.*, 1992; Tsaur *et al.*, 1998), and in Sfp-mediated traits, like sperm competition (Clark *et al.*, 1995, 2000; see Chapman, 2001, for review). Why is so much polymorphism maintained?

The answer may be that there is no single "most advantageous" allele for a particular Sfp. What is "best" may depend on the genotypes of the individuals involved in the interaction. Using sperm competition as an example, within the mated female a male's sperm must compete for fertilizations with sperm from other males with whom the female had mated or will mate. In the case of three different genotypes (A, B, and C), male A may outcompete male B, and male B may outcompete male C, but this does not mean that male A will outcompete male C (A > B, B > C, but C > A), when female genotype is held constant. This property of sperm competition, known as nontransitivity (Clark *et al.*, 2000), tells us that the outcome of a sperm competition interaction depends on the particular alleles of the competing males. Antagonistic pleiotropy may also lead to maintenance of polymorphism among Sfp genes (Fiumera *et al.*, 2005, 2007). For example, a polymorphism in the Sfp CG17331 is associated both with offensive sperm competitive ability and female refractoriness. When males carry a certain allele, they are better able to prevent their mates from remating, but are poorer at displacing a previous mate's sperm (Fiumera *et al.*, 2005). The dependence of female postmating behavior on the specific genotypic combinations of her multiple mates makes it difficult for one Sfp gene allele to "win."

The female's genotype, along with the interaction between male and female genotypes, also determines the outcome of sperm competition (Fig. 2.5; Clark and Begun, 1998; Clark *et al.*, 1999). Complicating matters further, there is no genetic correlation between male-derived and female-derived effects on sperm competition (Civetta and Clark, 2000a). When the genotypes of competing males are held constant, a change in female genotype can determine which male is the "winner." The complex interactions between the genotypes of

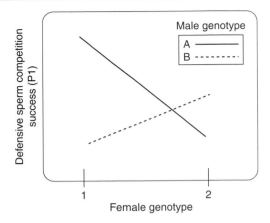

Figure 2.5. Male and female genotypes affect outcome of sperm competition experiments. The genotypes of the particular male and female involved in a mating affect the defensive component of sperm c, P1. P1 measures the proportion of offspring sired by the first male after the female has remated. In this hypothetical example, male A has a much higher P1 than male B when either of them mates with a female of genotype 1. However, when these same males mate a female of genotype 2, male B now has the advantage. Here, the second male to mate in each competition experiment does not vary. These interactions become even more complex when the genotype of the second male to mate is taken into consideration.

competing males and between male and female genotypes would make it nearly impossible for a particular male genotype to persevere in sperm competition. In fact, artificial selection even fails in the face of this complexity (Bjork *et al.*, 2007).

Males must compete against each other for fertilization success and against the female's sometimes conflicting interests. For example, one might expect males to benefit from inhibiting females from remating, whereas females may benefit from receiving the sperm of multiple males. It is also likely, given the differing magnitudes of investment in progeny production between males and females, that they have different optima for the level of postmating effects (such as ovulation rate) (Chippindale *et al.*, 2001). Conflict between the sexes due to these different optima has been interpreted to suggest a kind of evolutionary "arms race" (Fig. 2.6). First, male–male competition selects for alleles conferring higher male competitive reproductive success. These alleles may lower a female's lifetime progeny production for unknown reasons, resulting in selection for females to counteradapt to overcome harmful male alleles (Chippindale *et al.*, 2001; Civetta, 2003; Gavrilets *et al.*, 2001; Rice, 2000). At any given time in a population, this scenario of adaptation and counteradaptation would occur simultaneously for many traits, leading to high levels of allelic polymorphism at the loci involved.

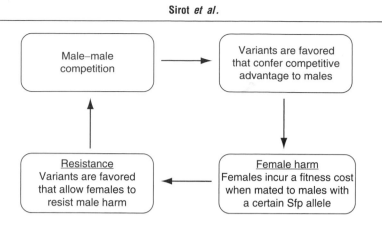

Figure 2.6. Sexual conflict may maintain polymorphisms in the population. Competition among males for mates/fertilizations drives selection for new alleles that confer a competitive advantage to males for some reproductive trait. For example, males with this allele may induce a higher rate of ovulation in females by way of a particular Sfp. This higher ovulation rate may be harmful to females due to higher energy costs, driving selection for alleles that allow females to resist the effects of the harmful male allele. This cycle continues, with males developing new competitive strategies, each with a potential female cost associated with them. As a result of this process occurring simultaneously at many loci, polymorphisms are maintained in the population. Adapted from Arnqvist and Rowe (2005).

Consistent with this prediction, we see that maintenance of variation in male-derived proteins depends on female genotype (reviewed in Chapman, 2001). Furthermore, variation in Sfp-mediated female postmating traits, such as female sperm usage patterns (determined by sperm competition experiments), is also dependent on female genotype (Clark, 2002; Clark et al., 1999; Lawniczak and Begun, 2005; McGraw et al., 2009).

        Experimental evolution studies support the hypothesis that evolution of Sfps is driven, in part, by sexual selection. When monogamy is enforced on *D. melanogaster* for multiple generations, males become less harmful and females become less resistant to harm than polygamous controls (Holland and Rice, 1999). In *Drosophila pseudoobscura* populations selected under monogamy, normal levels of promiscuity, or elevated levels of promiscuity, males differed in their ability to prevent remating of their mates (Crudgington et al., 2005). Counterintuitively, *D. pseudoobscura* males selected under monogamy induced greater refractory periods in females, suggesting that the males selected under promiscuous female conditions invested more in precopulatory competition than in postcopulatory competition. In another experiment, when female *D. melanogaster* were prevented from coevolving with a male population that was allowed to adapt to the static female genome, these males were able to induce higher remating rates in females and were better at the "defense" component of sperm competition (Rice, 1996). Though we do not know for certain whether any of these results are due to changes in Sfps (either at the sequence or expression levels), the Wigby *et al.* (2009) study described in Act 4 suggests that Sfps are involved in these types of adaptations because in this study changes in levels of transferred Sfps correlated with changes in postcopulatory traits in females (see also Bretman *et al.*, 2009).

A. Female contribution to molecular social interactions

Since many Sfps are rapidly evolving, we might expect parallel patterns of evolution for the many female proteins (yet to be identified, with the exception of SPR) with which they interact. In line with this prediction, there are rapidly evolving genes expressed in the female reproductive tract (Lawniczak and Begun, 2007; Prokupek *et al.*, 2008; Swanson *et al.*, 2004), although this group as a class is not enriched for rapidly evolving genes (Haerty *et al.*, 2007). However, molecular evidence for the type of "arms race" between the sexes, mediated by Sfp–female interactor pairs, is lacking. The only female interactor of an Sfp that has been identified, the G-protein coupled receptor SPR, has not undergone rapid evolution; in fact, it is strikingly conserved across evolutionarily distant lineages (Yapici *et al.*, 2008). Studies that have identified rapidly evolving female reproductive tract proteins found that they are enriched for serine proteases, not receptors as might have been predicted (e.g., Clark *et al.*, 2009; Swanson *et al.*, 2001), though several rapidly evolving female genes also have unknown functions. Some serine proteases expressed in female reproductive tracts are differentially expressed in response to mating (Lawniczak and Begun, 2007; Prokupek *et al.*, 2009), suggesting a possible direct or indirect interaction with Sfps.

Much work needs to be done to determine whether SPR's conservation is the norm in *Drosophila*, or an exception. Examples can be found from many other species where, unlike SP and SPR, variation in a male or female molecule seems to drive rapid evolution at its partner from the opposite sex. For instance, the abalone sperm lysin protein has undergone rapid diversification which is thought to be a response to sequence changes in its receptor on the egg (Clark *et al.*, 2009; Swanson and Vacquier, 1998; Swanson *et al.*, 2001). Before we can test the prediction that changes in male reproductive proteins cause changes to female molecules (or vice versa) and downstream behaviors, more male/female molecule pairs must be identified. Transcriptional profiling studies of females after mating (Lawniczak and Begun, 2004; Kapelnikov *et al.*, 2008b; Mack *et al.*, 2006; McGraw *et al.*, 2004, 2008), in conjunction with evolutionary studies will likely lead to identification of further promising candidates.

III. DISCUSSION AND FUTURE DIRECTIONS

We now know a great deal about how Sfps affect social interactions related to mating in *D. melanogaster*. However, the current body of research on Sfps is only the tip of the iceberg when it comes to understanding how the social interactions themselves affect Sfp usage and evolution. Future experiments can explore new social contexts in *D. melanogaster* for their effect on the overall amount of Sfps transferred, the effect on specific Sfps, and the ultimate reproductive consequences for each individual involved. Experimental evolution studies may test for changes in Sfp production or mating-induced transcriptional responses in response to altered population dynamics.

One of the greatest challenges to elucidating the relationship between organismal-level interactions and individual Sfps (and their female interactors) is the difficulty in performing experiments under "natural" conditions. Advantageous reproductive traits no doubt depend on environmental conditions. For example, the larval environment (McGraw *et al.*, 2007) and adult male nutrition (Fricke *et al.*, 2008) have been shown to affect postmating traits in *D. melanogaster*. In the lab, there are no predators and flies do not experience the dangers they might in nature, such as depletion of food supply, extreme temperatures, or desiccation. Simulating as wide an array of conditions as possible will provide valuable information on which social interactions are really important in nature, and how these interactions affect Sfp dynamics and postmating traits.

Several studies have considered at least one of the important conditions experienced in nature: genetic variation among individuals. While most mechanistic studies of Sfps have been carried out using standard lab strains, variation in sperm competition and other postmating traits have been linked to variation in male and female genotypes (Clark *et al.*, 1995, 1999; Fiumera *et al.*, 2006, 2007; McGraw *et al.*, 2009; Prout and Clark, 1996). Currently, the genomes of nearly 200 inbred strains derived from wild-caught *D. melanogaster* are being sequenced (Mackay *et al.*, 2008). This unprecedented genomic tool will provide a means for identifying the genetic basis of any number of complex traits, including behaviors affected by Sfps. Future experiments will be able to tease apart the effects of different genotypic combinations on social interactions such as receptivity to mating and sperm competition. From here, we can finally start to understand how females, not just males, act to control their reproductive success and to affect the reproductive success of their mates. Only then we will be able to comprehend how all levels of mating interactions come together to affect the complex molecular social interactions between *Drosophila* individuals.

Acknowledgments

We are grateful to Marla Sokolowski for the opportunity to contribute to this volume. We thank Nathan Clark, Geoff Findlay, Lisa McGraw, and Alex Wong for comments on an earlier version of this paper. We are grateful for support from NIH/NICHD grant R01-HD038921 (to MFW), from an NSF predoctoral fellowship to BAL, and for NIH training grant T32-GM07617 (in Genetics & Development) for support of JLS, and training grant T32-HD052471 (in Reproductive Genomics) and grant R01-HD059060 (to MFW and A. Clark) for support of CYC.

References

Adams, E. M., and Wolfner, M. F. (2007). Seminal proteins but not sperm induce morphological changes in the *Drosophila melanogaster* female reproductive tract during sperm storage. *J. Insect Physiol.* **53,** 319–331.

Adkins-Regan, E. (2009). Hormones and sexual differentiation of avian social behavior. *Dev. Neurosci.* **31,** 342–350.

Aguadé, M. (1998). Different forces drive the evolution of the Acp26Aa and Acp26Ab accessory gland genes in the *Drosophila melanogaster* species complex. *Genetics* **150,** 1079–1089.

Aguadé, M., Miyashita, N., and Langley, C. H. (1992). Polymorphism and divergence in the mst26a male accessory gland gene region in *Drosophila. Genetics* **132,** 755–770.

Aigaki, T., Fleischmann, I., Chen, P. S., and Kubli, E. (1991). Ectopic expression of sex peptide alters reproductive behavior of female *Drosophila melanogaster. Neuron* **7,** 557–563.

Allen, A. K., and Spradling, A. C. (2008). The Sf1-related nuclear hormone receptor Hr39 regulates *Drosophila* female reproductive tract development and function. *Development* **135,** 311–321.

Anderson, R. C. (1945). A study of the factors affecting fertility of lozenge females of *Drosophila melanogaster. Genetics* **30,** 280–296.

Andolfatto, P. (2005). Adaptive evolution of non-coding DNA in *Drosophila. Nature* **437,** 1149–1152.

Arbeitman, M. N., Furlong, E. E. M., Imam, F., Johnson, E., Null, B. H., Baker, B. S., Krasnow, M. A., Scott, M. P., Davis, R. W., and White, K. P. (2002). Gene expression during the life cycle of *Drosophila melanogaster. Science* **297,** 2270–2275.

Arnqvist, G., and Nilsson, T. (2000). The evolution of polyandry: multiple mating and female fitness in insects. *Anim. Behav.* **60,** 145–164.

Arnqvist, G., and Rowe, L. (2005). Sexual conflict. Princeton University Press, Princeton.

Avila, F. W., and Wolfner, M. F. (2009). Acp36DE is required for uterine conformational changes in mated *Drosophila* females. *Proc. Natl. Acad. Sci. USA* **106**(37), 15796–15800.

Barnes, A. I., Wigby, S., Boone, J. M., Partridge, L., and Chapman, T. (2008). Feeding, fecundity and lifespan in female *Drosophila melanogaster. Proc. R. Soc. B-Biol. Sci.* **275,** 1675–1683.

Bartoszewski, S., Luschnig, S., Desjeux, I., Grosshans, J., and Nusslein-Volhard, C. (2004). *Drosophila* p24 homologues eclair and baiser are necessary for the activity of the maternally expressed Tkv receptor during early embryogenesis. *Mech. Dev.* **121,** 1259–1273.

Begun, D. J., and Lindfors, H. A. (2005). Rapid evolution of genomic Acp complement in the melanogaster subgroup of *Drosophila. Mol. Biol. Evol.* **22,** 2010–2021.

Bertram, M. J., Neubaum, D. M., and Wolfner, M. F. (1996). Localization of the *Drosophila* male accessory gland protein Acp36DE in the mated female suggests a role in sperm storage. *Insect Biochem. Mol. Biol.* **26,** 971–980.

Birkhead, T. R., and Møller, A. P. (1998). Sperm competition and sexual selection. Academic Press, San Diego.

Bjork, A., Starmer, W. T., Higginson, D. M., Rhodes, C. J., and Pitnick, S. (2007). Complex interactions with females and rival males limit the evolution of sperm offence and defence. *Proc. R. Soc. Lond. B Biol.* **274,** 1779–1788.

Bloch Qazi, M. C., and Wolfner, M. F. (2003). An early role for the *Drosophila melanogaster* male seminal protein Acp36DE in female sperm storage. *J. Exp. Biol.* **206,** 3521–3528.

Bloch Qazi, M. C., Heifetz, Y., and Wolfner, M. F. (2003). The developments between gametogenesis and fertilization: Ovulation and female sperm storage in *Drosophila melanogaster. Dev. Biol.* **256,** 195–211.

Boltz, K. A., Ellis, L. L., and Carney, G. E. (2007). *Drosophila melanogaster* p24 genes have developmental, tissue-specific, and sex-specific expression patterns and functions. *Dev. Dyn.* **236,** 544–555.

Boorman, E., and Parker, G. A. (1976). Sperm (ejaculate) competition in *Drosophila melanogaster,* and reproductive value of females to males in relation to female age and mating status. *Ecol. Entomol.* **1,** 145–155.

Boulétreau-Merle, J. (1977). Role of spermathecae in sperm utilization and stimulation of oogenesis in *Drosophila melanogaster. J. Insect Physiol.* **23,** 1099–1104.

Bretman, A., Fricke, C., and Chapman, T. (2009). Plastic responses of male *Drosophila melanogaster* to the level of sperm competition increase male reproductive fitness. *Proc. R. Soc. Lond. B Biol.* **276** (1662), 1705–1711.

Bretman, A., Lawniczak, M. K. N., Boone, J. M., and Chapman, T. A mating plug protein reduces early female remating in *Drosophila melanogaster. J. Insect Physiol.* (in press). doi:10.1016/j.jinsphys.2009.09.010.

Byrne, P. G., and Rice, W. R. (2006). Evidence for adaptive male mate choice in the fruit fly *Drosophila melanogaster. Proc. R. Soc. Lond. B Biol.* **273,** 917–922.

Cameron, E., Day, T., and Rowe, L. (2007). Sperm competition and the evolution of ejaculate composition. *Am. Nat.* **169,** E158–E172.

Carney, G. E., and Taylor, B. J. (2003). logjam encodes a predicted EMP24/GP25 protein that is required for *Drosophila* oviposition behavior. *Genetics* **164,** 173–186.

Carvalho, G. B., Kapahi, P., Anderson, D. J., and Benzer, S. (2006). Allocrine modulation of feeding behavior by the sex peptide of *Drosophila. Curr. Biol.* **16,** 692–696.

Chapman, T. (2001). Seminal fluid-mediated fitness traits in *Drosophila. Heredity* **87,** 511–521.

Chapman, T. (2008). The soup in my fly: Evolution, form and function of seminal fluid proteins. *PLoS Biol.* **6,** 1379–1382.

Chapman, T., and Davies, S. J. (2004). Functions and analysis of the seminal fluid proteins of male *Drosophila melanogaster* fruit flies. *Peptides* **25,** 1477–1490.

Chapman, T., Liddle, L. F., Kalb, J. M., Wolfner, M. F., and Partridge, L. (1995). Cost of mating in *Drosophila melanogaster* females is mediated by male accessory gland products. *Nature* **373,** 241–244.

Chapman, T., Neubaum, D. M., Wolfner, M. F., and Partridge, L. (2000). The role of male accessory gland protein Acp36DE in sperm competition in *Drosophila melanogaster. Proc. R. Soc. Lond. B Biol.* **267,** 1097–1105.

Chapman, T., Herndon, L. A., Heifetz, Y., Partridge, L., and Wolfner, M. F. (2001). The Acp26Aa seminal fluid protein is a modulator of early egg hatchability in *Drosophila melanogaster. Proc. R. Soc. Lond. B Biol.* **268,** 1647–1654.

Chapman, T., Bangham, J., Vinti, G., Seifried, B., Lung, O., Wolfner, M. F., Smith, H. K., and Partridge, L. (2003). The sex peptide of *Drosophila melanogaster*: Female post-mating responses analyzed by using RNA interference. *Proc. Natl. Acad. Sci. USA* **100,** 9923–9928.

Chen, J. S., and Raikhel, A. S. (1996). Subunit cleavage of mosquito pro-vitellogenin by a subtilisin-like convertase. *Proc. Natl. Acad. Sci. USA* **93,** 6186–6190.

Chen, P. S., Stummzollinger, E., Aigaki, T., Balmer, J., Bienz, M., and Bohlen, P. (1988). A male accessory gland peptide that regulates reproductive behavior of female *Drosophila melanogaster. Cell* **54,** 291–298.

Chintapalli, V. R., Wang, J., and Dow, J. A. T. (2007). Using FlyAtlas to identify better *Drosophila melanogaster* models of human disease. *Nat. Genet.* **39,** 715–720.

Chippindale, A. K., Gibson, J. R., and Rice, W. R. (2001). Negative genetic correlation for adult fitness between sexes reveals ontogenetic conflict in *Drosophila*. *Proc. Natl. Acad. Sci. USA* **98,** 1671–1675.

Christiansen, A. E., Keisman, E. L., Ahmad, S. M., and Baker, B. S. (2002). Sex comes in from the cold: The integration of sex and pattern. *Trends Genet.* **18,** 510–516.

Civetta, A. (2003). Shall we dance or shall we fight? Using DNA sequence data to untangle controversies surrounding sexual selection. *Genome* **46,** 925–929.

Civetta, A., and Clark, A. G. (2000a). Chromosomal effects on male and female components of sperm precedence in *Drosophila*. *Genet. Res.* **75,** 143–151.

Civetta, A., and Clark, A. G. (2000b). Correlated effects of sperm competition and postmating female mortality. *Proc. Natl. Acad. Sci. USA* **97,** 13162–13165.

Clark, A. G. (2002). Sperm competition and the maintenance of polymorphism. *Heredity* **88,** 148–153.

Clark, A. G., and Begun, D. J. (1998). Female genotypes affect sperm displacement in *Drosophila*. *Genetics* **149,** 1487–1493.

Clark, A. G., Aguadé, M., Prout, T., Harshman, L. G., and Langley, C. H. (1995). Variation in sperm displacement and its association with accessory gland protein loci in *Drosophila melanogaster*. *Genetics* **139,** 189–201.

Clark, A. G., Begun, D. J., and Prout, T. (1999). Female x male interactions in *Drosophila* sperm competition. *Science* **283,** 217–220.

Clark, A. G., Dermitzakis, E. T., and Civetta, A. (2000). Nontransitivity of sperm precedence in *Drosophila*. *Evolution* **54,** 1030–1035.

Clark, N. L., Aagaard, J. E., and Swanson, W. J. (2006). Evolution of reproductive proteins from animals and plants. *Reproduction* **131,** 11–22.

Clark, N. L., Gasper, J., Sekino, M., Springer, S. A., Aquadro, C. F., and Swanson, W. J. (2009). Coevolution of interacting fertilization proteins. *PLoS Genet.* **5,** e1000570.

Cline, T. W., and Meyer, B. J. (1996). Vive la différence: Males vs females in flies vs worms. *Annu. Rev. Genet.* **30,** 637–702.

Cole, S. H., Carney, G. E., McClung, C. A., Willard, S. S., Taylor, B. J., and Hirsh, J. (2005). Two functional but noncomplementing *Drosophila* tyrosine decarboxylase genes. *J. Biol. Chem.* **280,** 14948–14955.

Cordts, R., and Partridge, L. (1996). Courtship reduces longevity of male *Drosophila melanogaster*. *Anim. Behav.* **52,** 269–278.

Crudgington, H. S., Beckerman, A. P., Brüstle, L., Green, K., and Snook, R. R. (2005). Experimental removal and elevation of sexual selection: Does sexual selection generate manipulative males and resistant females? *Am. Nat.* **165**(Suppl. 5), S72–S87.

Dean, M. D., Good, J. M., and Nachman, M. W. (2008). Adaptive evolution of proteins secreted during sperm maturation: An analysis of the mouse epididymal transcriptome. *Mol. Biol. Evol.* **25,** 383–392.

de Bono, M., and Maricq, A. V. (2005). Neuronal substrates of complex behaviors in C. *elegans*. *Ann. Rev. Neurosci.* **28,** 451–501.

DiBenedetto, A. J., Harada, H. A., and Wolfner, M. F. (1990). Structure, cell-specific expression, and mating-induced regulation of a *Drosophila melanogaster* male accessory gland gene. *Dev. Biol.* **139,** 134–148.

Dickson, B. J. (2008). Wired for sex: The neurobiology of *Drosophila* mating decisions. *Science* **322,** 904–909.

Domanitskaya, E. V., Liu, H. F., Chen, S. J., and Kubli, E. (2007). The hydroxyproline motif of male sex peptide elicits the innate immune response in *Drosophila* females. *FEBS J.* **274,** 5659–5668.

Eberhard, W. G. (1996). Female control: Sexual selection by cryptic female choice. Princeton University Press, Princeton.

Ejima, A., Smith, B. P. C., Lucas, C., Van Naters, W. V., Miller, C. J., Carlson, J. R., Levine, J. D., and Griffith, L. C. (2007). Generalization of courtship learning in *Drosophila* is mediated by *cis*-vaccenyl acetate. *Curr. Biol.* **17,** 599–605.

Ellison, P. T., and Gray, P. B. (eds.) (2009). Enodcrinology of Social Relationships. Harvard University Press, Cambridge, MA.

Fay, J. C., and Wu, C. I. (2000). Hitchhiking under positive Darwinian selection. *Genetics* **155,** 1405–1413.

Fedorka, K. M., Linder, J. E., Winterhalter, W., and Promislow, D. (2007). Post-mating disparity between potential and realized immune response in *Drosophila melanogaster*. *Proc. R. Soc. Lond. B Biol.* **274,** 1211–1217.

Ferveur, J. F. (1997). The pheromonal role of cuticular hydrocarbons in *Drosophila melanogaster*. *Bioessays* **19,** 353–358.

Findlay, G. D., Yi, X. H., MacCoss, M. J., and Swanson, W. J. (2008). Proteomics reveals novel *Drosophila* seminal fluid proteins transferred at mating. *PLoS Biol.* **6,** 1417–1426.

Findlay, G. D., MacCoss, M. J., and Swanson, W. J. (2009). Proteomic discovery of previously unannotated, rapidly evolving seminal fluid genes in *Drosophila*. *Genome Res.* **19,** 886–896.

Fiumera, A. C., Dumont, B. L., and Clark, A. G. (2005). Sperm competitive ability in *Drosophila melanogaster* associated with variation in male reproductive proteins. *Genetics* **169,** 243–257.

Fiumera, A. C., Dumont, B. L., and Clark, A. G. (2006). Natural variation in male-induced 'cost-of-mating' and allele-specific association with male reproductive genes in *Drosophila melanogaster*. *Philos. Trans. R. Soc. Lond. B Biol. Sci.* **361,** 355–361.

Fiumera, A. C., Dumont, B. L., and Clark, A. G. (2007). Associations between sperm competition and natural variation in male reproductive genes on the third chromosome of *Drosophila melanogaster*. *Genetics* **176,** 1245–1260.

Fowler, G. L. (1973). Some aspects of reproductive biology of *Drosophila*—Sperm transfer, sperm storage, and sperm utilization. *Adv. Genet.* **17,** 293–360.

Fowler, K., and Partridge, L. (1989). A cost of mating in female fruit flies. *Nature* **338,** 760–761.

Friberg, U. (2006). Male perception of female mating status: Its effect on copulation duration, sperm defence and female fitness. *Anim. Behav.* **72,** 1259–1268.

Fricke, C., Bretman, A., and Chapman, T. (2008). Adult male nutrition and reproductive success in *Drosophila melanogaster*. *Evolution* **62,** 3170–3177.

Gao, Z. Q., Ruden, D. M., and Lu, X. Y. (2003). PKD2 cation channel is required for directional sperm movement and male fertility. *Curr. Biol.* **13,** 2175–2178.

Garden, R. W., Shippy, S. A., Li, L. J., Moroz, T. P., and Sweedler, J. V. (1998). Proteolytic processing of the *Aplysia* egg-laying hormone prohormone. *Proc. Natl. Acad. Sci. USA* **95,** 3972–3977.

Gavrilets, S., Arnqvist, G., and Friberg, U. (2001). The evolution of female mate choice by sexual conflict. *Proc. R. Soc. Lond. B Biol.* **268,** 531–539.

Gilbert, D. G., Richmond, R. C., and Sheehan, K. B. (1981). Studies of esterase 6 in *Drosophila melanogaster*. 5. Progeny production and sperm use in females inseminated by males having active or null alleles. *Evolution* **35,** 21–37.

Gilchrist, A. S., and Partridge, L. (2000). Why it is difficult to model sperm displacement in *Drosophila melanogaster*: The relation between sperm transfer and copulation duration. *Evolution* **54,** 534–542.

Gillott, C. (2003). Male accessory gland secretions: Modulators of female reproductive physiology and behavior. *Annu. Rev. Entomol.* **48,** 163–184.

Greenspan, R. J., and Ferveur, J. F. (2000). Courtship in *Drosophila*. *Annu. Rev. Genet.* **34,** 205–232.

Haerty, W., Jagadeeshan, S., Kulathinal, R. J., Wong, A., Ravi Ram, K., Sirot, L. K., Levesque, L., Artieri, C. G., Wolfner, M. F., Civetta, A., and Singh, R. S. (2007). Evolution in the fast lane: Rapidly evolving sex-related genes in *Drosophila*. *Genetics* **177**, 1321–1335.

Hall, J. C. (1994). The mating of a fly. *Science* **264**, 1702–1714.

Harshman, L. G., and Clark, A. G. (1998). Inference of sperm competition from broods of field-caught *Drosophila*. *Evolution* **52**, 1334–1341.

Hasemeyer, M., Yapici, N., Heberlein, U., and Dickson, B. J. (2009). Sensory neurons in the *Drosophila* genital tract regulate female reproductive behavior. *Neuron* **61**, 511–518.

Heifetz, Y., and Wolfner, M. F. (2004). Mating, seminal fluid components, and sperm cause changes in vesicle release in the *Drosophila* female reproductive tract. *Proc. Natl. Acad. Sci. USA* **101**, 6261–6266.

Heifetz, Y., Lung, O., Frongillo, E. A., and Wolfner, M. F. (2000). The *Drosophila* seminal fluid protein Acp26Aa stimulates release of oocytes by the ovary. *Curr. Biol.* **10**, 99–102.

Heifetz, Y., Tram, U., and Wolfner, M. F. (2001). Male contributions to egg production: the role of accessory gland products and sperm in *Drosophila melanogaster*. *Proc. R. Soc. Lond. B. Biol.* **268**, 175–180.

Heifetz, Y., Vandenberg, L. N., Cohn, H. I., and Wolfner, M. F. (2005). Two cleavage products of the *Drosophila* accessory gland protein ovulin can independently induce ovulation. *Proc. Natl. Acad. Sci. USA* **102**, 743–748.

Herndon, L. A., and Wolfner, M. F. (1995). A *Drosophila* seminal fluid protein, Acp26Aa, stimulates egg-laying in females for 1 day after mating. *Proc. Natl. Acad. Sci. USA* **92**, 10114–10118.

Hihara, F. (1981). Effects of the male accessory gland secretion on oviposition and remating in females of *Drosophila melanogaster*. *Zool. Mag. (Tokyo)* **90**, 307–316.

Holland, B., and Rice, W. R. (1999). Experimental removal of sexual selection reverses intersexual antagonistic coevolution and removes a reproductive load. *Proc. Natl. Acad. Sci. USA* **96**, 5083–5088.

Holman, L. (2009). *Drosophila melanogaster* seminal fluid can protect the sperm of other males. *Funct. Ecol.* **23**, 180–186.

Hook, V. Y. H., Azaryan, A. V., Hwang, S. R., and Tezapsidis, N. (1994). Proteases and the emerging role of protease inhibitors in prohormone processing. *FASEB J.* **8**, 1269–1278.

Iida, K., and Cavener, D. R. (2004). Glucose dehydrogenase is required for normal sperm storage and utilization in female *Drosophila melanogaster*. *J. Exp. Biol.* **207**, 675–681.

Imhof, M., Harr, B., Brem, G., and Schlötterer, C. (1998). Multiple mating in wild *Drosophila melanogaster* revisited by microsatellite analysis. *Mol. Ecol.* **7**, 915–917.

Innocenti, P., and Morrow, E. H. (2009). Immunogenic males: A genome-wide analysis of reproduction and the cost of mating in *Drosophila melanogaster* females. *J. Evol. Biol.* **22**, 964–973.

Kalb, J. M., DiBenedetto, A. J., and Wolfner, M. F. (1993). Probing the function of *Drosophila melanogaster* accessory glands by directed cell ablation. *Proc. Natl. Acad. Sci. USA* **90**, 8093–8097.

Kapelnikov, A., Zelinger, E., Gottlieb, Y., Rhrissorrakrai, K., Gunsalus, K. C., and Heifetz, Y. (2008b). Mating induces an immune response and developmental switch in the *Drosophila* oviduct. *Proc. Natl. Acad. Sci. USA* **105**, 13912–13917.

Kapelnikov, A., Rivlin, P. K., Hoy, R. R., and Heifetz, Y. (2008a). Tissue remodeling: A mating-induced differentiation program for the *Drosophila* oviduct. *BMC Dev. Biol.* **8**, 114.

Kaufmann, P., and Demerec, M. (1942). Utilization of sperm by the female *Drosophila melanogaster*. *Am. Nat.* **76**, 445–469.

Kelleher, E., Swanson, W. J., and Markow, T. A. (2007). Gene duplication and adaptive evolution of digestive proteases in *Drosophila arizonae* female reproductive tracts. *PLoS Genet.* **3**, e148.

Kern, A. D., Jones, C. D., and Begun, D. J. (2004). Molecular population genetics of male accessory gland proteins in the *Drosophila simulans* complex. *Genetics* **167**, 725–735.

Keverne, E. B., and Curley, J. P. (2004). Vasopressin, oxytocin and social behaviour. *Curr. Opin. Neurobiol.* **14,** 777–783.

Kubli, E. (2003). Sex-peptides: Seminal peptides of the *Drosophila* male. *Cell. Mol. Life Sci.* **60,** 1689–1704.

Lange, A. B. (2009). Tyramine: From octopamine precursor to neuroactive chemical in insects. *Gen. Comp. Endocrinol.* **162,** 18–26.

Lawniczak, M. K. N., and Begun, D. J. (2004). A genome-wide analysis of courting and mating responses in *Drosophila melanogaster* females. *Genome* **47,** 900–910.

Lawniczak, M. K. N., and Begun, D. J. (2005). A QTL analysis of female variation contributing to refractoriness and sperm competition in *Drosophila melanogaster*. *Genet. Res.* **86,** 107–114.

Lawniczak, M. K. N., and Begun, D. J. (2007). Molecular population genetics of female-expressed mating-induced serine proteases in *Drosophila melanogaster*. *Mol. Biol. Evol.* **24,** 1944–1951.

Lee, H. G., Seong, C. S., Kim, Y. C., Davis, R. L., and Han, K. A. (2003). Octopamine receptor OAMB is required for ovulation in *Drosophila melanogaster*. *Dev. Biol.* **264,** 179–190.

Lee, H. G., Rohila, S., and Han, K. A. (2009). The octopamine receptor OAMB mediates ovulation via Ca2+/calmodulin-dependent protein kinase II in the *Drosophila* oviduct epithelium. *PLoS One* **4,** e4716.

Lefevre, G., and Jonsson, U. B. (1962). Sperm transfer, storage, displacement, and utilization in *Drosophila melanogaster*. *Genetics* **47,** 1719–1736.

LeFranc, A., and Bundgaard, J. (2000). The influence of male and female body size on copulation duration and fecundity in *Drosophila melanogaster*. *Hereditas* **132,** 243–247.

Linklater, J. R., Wertheim, B., Wigby, S., and Chapman, T. (2007). Ejaculate depletion patterns evolve in response to experimental manipulation of sex ratio in *Drosophila melanogaster*. *Evolution* **61,** 2027–2034.

Liu, H. F., and Kubli, E. (2003). Sex-peptide is the molecular basis of the sperm effect in *Drosophila melanogaster*. *Proc. Natl. Acad. Sci. USA* **100,** 9929–9933.

Long, T. A. F., Stewart, A. D., and Miller, P. M. (2009). Potential confounds to an assay of cross-generational fitness benefits of mating and male seminal fluid. *Biol. Lett.* **5,** 26–27.

Ludwig, M. Z., Uspensky, I. I., Ivanov, A. I., Kopantseva, M. R., Dianov, C. M., Tamarina, N. A., and Korochkin, L. I. (1991). Genetic control and expression of the major ejaculatory bulb protein (Peb-ME) in *Drosophila melanogaster*. *Biochem. Genet.* **29,** 215–239.

Lung, O., and Wolfner, M. F. (1999). *Drosophila* seminal fluid proteins enter the circulatory system of the mated female fly by crossing the posterior vaginal wall. *Insect Biochem. Mol. Biol.* **29,** 1043–1052.

Lung, O., and Wolfner, M. F. (2001). Identification and characterization of the major *Drosophila melanogaster* mating plug protein. *Insect Biochem. Mol. Biol.* **31,** 543–551.

Lung, O., Kuo, L., and Wolfner, M. F. (2001). *Drosophila* males transfer antibacterial proteins from their accessory gland and ejaculatory duct to their mates. *J. Insect Physiol.* **47,** 617–622.

Mack, P. D., Kapelnikov, A., Heifetz, Y., and Bender, M. (2006). Mating-responsive genes in reproductive tissues of female *Drosophila melanogaster*. *Proc. Natl. Acad. Sci. USA* **103,** 10358–10363.

Mackay, T., Richards, S., and Gibbs, R. (2008). Proposal to sequence a *Drosophila* genetic reference panel: A community resource for the study of genotypic and phenotypic variation http://www.genome.gov/Pages/Research/Sequencing/SeqProposals/DrosophilaGeneticReferencePanelWhitepaper4_9_08_final_combined.pdf.

Manning, A. (1962). Sperm factor affecting receptivity of *Drosophila melanogaster* females. *Nature* **194,** 252–253.

Manning, A. (1967). Control of sexual receptivity in female *Drosophila*. *Anim. Behav.* **15,** 239–250.

Manoli, D. S., Meissner, G. W., and Baker, B. S. (2006). Blueprints for behavior: Genetic specification of neural circuitry for innate behaviors. *Trends Neurosci.* **29,** 444–451.

Marcillac, F., and Ferveur, J. F. (2004). A set of female pheromones affects reproduction before, during and after mating in *Drosophila*. *J. Exp. Biol.* **207**, 3927–3933.

Markow, T. A., and O'Grady, P. M. (2005). Evolutionary genetics of reproductive behavior in *Drosophila*: Connecting the dots. *Annu. Rev. Genet.* **39**, 263–291.

Mazzi, D., Kesaniemi, J., Hoikkala, A., and Klappert, K. (2009). Sexual conflict over the duration of copulation in *Drosophila montana*: Why is longer better? *BMC Evol. Biol.* **9**, 132.

McGraw, L. A., Gibson, G., Clark, A. G., and Wolfner, M. F. (2004). Genes regulated by mating, sperm, or seminal proteins in mated female *Drosophila melanogaster*. *Curr. Biol.* **14**, 1509–1514.

McGraw, L. A., Clark, A. G., and Wolfner, M. F. (2008). Post-mating gene expression profiles of female *Drosophila melanogaster* in response to time and to four male accessory gland proteins. *Genetics* **179**, 1395–1408.

McGraw, L. A., Gibson, G., Clark, A. G., and Wolfner, M. F. (2009). Strain-dependent differences in several reproductive traits are not accompanied by early postmating transcriptome changes in female *Drosophila melanogaster*. *Genetics* **181**, 1273–1280.

Meikle, D. B., Sheehan, K. B., Phillis, D. M., and Richmond, R. C. (1990). Localization and longevity of seminal-fluid esterase-6 in mated female *Drosophila melanogaster*. *J. Insect Physiol.* **36**, 93–101.

Middleton, C. A., Nongthomba, U., Parry, K., Sweeney, S. T., Sparrow, J. C., and Elliott, C. J. H. (2006). Neuromuscular organization and aminergic modulation of contractions in the *Drosophila* ovary. *BMC Biol.* **4**, 17.

Monastirioti, M. (1999). Biogenic amine systems in the fruit fly *Drosophila melanogaster*. *Microsc. Res. Tech.* **45**, 106–121.

Monastirioti, M. (2003). Distinct octopamine cell population residing in the CNS abdominal ganglion controls ovulation in *Drosophila melanogaster*. *Dev. Biol.* **264**, 38–49.

Monsma, S. A., and Wolfner, M. F. (1988). Structure and expression of a *Drosophila* male accessory gland gene whose product resembles a peptide pheromone precursor. *Genes Dev.* **2**, 1063–1073.

Monsma, S. A., Harada, H. A., and Wolfner, M. F. (1990). Synthesis of 2 *Drosophila* male accessory gland proteins and their fate after transfer to the female during mating. *Dev. Biol.* **142**, 465–475.

Moshitzky, P., Fleischmann, I., Chaimov, N., Saudan, P., Klauser, S., Kubli, E., and Applebaum, S. W. (1996). Sex peptide activates juvenile hormone biosynthesis in the *Drosophila melanogaster* corpus allatum. *Arch. Insect Biochem.* **32**, 363–374.

Mueller, J. L., Ripoll, D. R., Aquadro, C. F., and Wolfner, M. F. (2004). Comparative structural modeling and inference of conserved protein classes in *Drosophila* seminal fluid. *Proc. Natl. Acad. Sci. USA* **101**, 13542–13547.

Mueller, J. L., Ravi Ram, K., McGraw, L. A., Bloch Qazi, M. C., Siggia, E. D., Clark, A. G., Aquadro, C., and Wolfner, M. F. (2005). Cross-species comparison of *Drosophila* male accessory gland protein genes. *Genetics* **171**, 131–143.

Mueller, J. L., Linklater, J. R., Ravi Ram, K., Chapman, T., and Wolfner, M. R. (2008). Targeted gene deletion and phenotypic analysis of the *Drosophila melanogaster* seminal fluid protease inhibitor Acp62F. *Genetics* **178**, 1605–1614.

Neubaum, D. M., and Wolfner, M. F. (1999). Mated *Drosophila melanogaster* females require a seminal fluid protein, Acp36DE, to store sperm efficiently. *Genetics* **153**, 845–857.

Panhuis, T. M., Clark, N. L., and Swanson, W. J. (2006). Rapid evolution of reproductive proteins in abalone and *Drosophila*. *Philos. Trans. R. Soc. Lond. B Biol. Sci.* **361**, 261–268.

Park, M., and Wolfner, M. F. (1995). Male and female cooperate in the prohormone-like processing of a *Drosophila melanogaster* seminal fluid protein. *Dev. Biol.* **171**, 694–702.

Pelosi, P., Calvello, M., and Ban, L. P. (2005). Diversity of odorant-binding proteins and chemo-sensory proteins in insects. *Chem. Sens.* **30**, i291–i292.

Peng, J., Chen, S., Busser, S., Liu, H. F., Honegger, T., and Kubli, E. (2005a). Gradual release of sperm bound sex-peptide controls female postmating behavior in *Drosophila*. *Curr. Biol.* **15**, 207–213.

Peng, J., Zipperlen, P., and Kubli, E. (2005b). *Drosophila* sex peptide stimulates female innate immune system after mating via the Toll and Imd pathways. *Curr. Biol.* **15,** 1690–1694.

Pilpel, N., Nezer, I., Applebaum, S. W., and Heifetz, Y. (2008). Mating-increases trypsin in female *Drosophila* hemolymph. *Insect Biochem. Mol. Biol.* **38,** 320–330.

Pitnick, S., Spicer, G. S., and Markow, T. A. (1995). How long is a giant sperm. *Nature* **375,** 109.

Poiani, A. (2006). Complexity of seminal fluid: A review. *Behav. Ecol. Sociobiol.* **60,** 289–310.

Priest, N. K., Galloway, L. F., and Roach, D. A. (2008a). Mating frequency and inclusive fitness in *Drosophila melanogaster. Am. Nat.* **171,** 10–21.

Priest, N. K., Roach, D. A., and Galloway, L. F. (2008b). Cross-generational fitness benefits of mating and male seminal fluid. *Biol. Lett.* **4,** 6–8.

Priest, N. K., Roach, D. A., and Galloway, L. F. (2009). Support for a pluralistic view of behavioural evolution. *Biol. Lett.* **5,** 28–29.

Prokupek, A., Hoffmann, F., Eyun, S. I., Moriyama, E., Zhou, M., and Harshman, L. (2008). An evolutionary expressed sequence tag analysis of *Drosophila* spermatheca genes. *Evolution* **62,** 2936–2947.

Prokupek, A. M., Kachman, S. D., Ladunga, I., and Harshman, L. G. (2009). Transcriptional profiling of the sperm storage organs of *Drosophila melanogaster. Insect Mol. Biol.* **18,** 465–475.

Prout, T., and Clark, A. G. (1996). Polymorphism in genes that influence sperm displacement. *Genetics* **144,** 401–408.

Prout, T., and Clark, A. G. (2000). Seminal fluid causes temporarily reduced egg hatch in previously mated females. *Proc. R. Soc. Lond. B Biol.* **267,** 201–203.

Ravi Ram, K., and Wolfner, M. F. (2007a). Seminal influences: *Drosophila* Acps and the molecular interplay between males and females during reproduction. *Integr. Comp. Biol.* **47,** 427–445.

Ravi Ram, K., and Wolfner, M. F. (2007b). Sustained post-mating response in *Drosophila melanogaster* requires multiple seminal fluid proteins. *PLoS Genet.* **3,** 2428–2438.

Ravi Ram, K., and Wolfner, M. F. (2009). A network of interactions among seminal proteins underlies the long-term postmating response in *Drosophila. Proc. Natl. Acad. Sci. USA* **106,** 15384–15389.

Ravi Ram, K., Ji, S., and Wolfner, M. F. (2005). Fates and targets of male accessory gland proteins in mated female *Drosophila melanogaster. Insect Biochem. Mol. Biol.* **35,** 1059–1071.

Ravi Ram, K., Sirot, L. K., and Wolfner, M. F. (2006). Predicted seminal astacin-like protease is required for processing of reproductive proteins in *Drosophila melanogaster. Proc. Natl. Acad. Sci. USA* **103,** 18674–18679.

Rice, W. R. (1996). Sexually antagonistic male adaptation triggered by experimental arrest of female evolution. *Nature* **381,** 232–234.

Rice, W. R. (2000). Dangerous liaisons. *Proc. Natl. Acad. Sci. USA* **97,** 12953–12955.

Ringo, J. (1996). Sexual receptivity in insects. *Annu. Rev. Entomol.* **41,** 473–494.

Rodríguez-Valentín, R., Lopez-Gonzalez, I., Jorquera, R., Laarca, P., Zurita, M., and Reynaud, E. (2006). Oviduct contraction in *Drosophila* is modulated by a neural network that is both octopaminergic and glutamatergic. *J. Cell. Physiol.* **209,** 183–198.

Samakovlis, C., Kylsten, P., Kimbrell, D. A., Engstrom, A., and Hultmark, D. (1991). The andropin gene and its product, a male-specific antibacterial peptide in *Drosophila melanogaster. EMBO J.* **10,** 163–169.

Schully, S. D., and Hellberg, M. E. (2006). Positive selection on nucleotide substitutions and indels in accessory gland proteins of the *Drosophila pseudoobscura* subgroup. *J. Mol. Evol.* **62,** 793–802.

Scott, D. (1986). Sexual mimicry regulates the attractiveness of mated *Drosophila melanogaster* females. *Proc. Natl. Acad. Sci. USA* **83,** 8429–8433.

Shirangi, T. R., and McKeown, M. (2007). Sex in flies: What 'body-mind' dichotomy? *Dev. Biol.* **306,** 10–19.

Simmons, L. W. (2001). Sperm competition and its evolutionary consequences in the insects. Princeton University Press, Princeton.

Singh, S. R., and Singh, B. N. (2004). Female remating in *Drosophila*: Comparison of duration of copulation between first and second matings in six species. *Curr. Sci.* **86,** 465–470.

Sirot, L. K., Buehner, N. A., Fiumera, A. C., and Wolfner, M. F. (2009). Seminal fluid protein depletion and replenishment in the fruit fly, *Drosophila melanogaster*: An ELISA-based method for tracking individual ejaculates. *Behav. Ecol. Sociobiol.* **63,** 1505–1513.

Smith, A. A., Holldobler, B., and Liebig, J. (2008). Hydrocarbon signals explain the pattern of worker and egg policing in the ant *Aphaenogaster cockerelli. J. Chem. Ecol.* **34,** 1275–1282.

Soller, M., Bownes, M., and Kubli, E. (1999). Control of oocyte maturation in sexually mature *Drosophila* females. *Dev. Biol.* **208,** 337–351.

Soller, M., Haussmann, I. U., Hollmann, M., Choffat, Y., White, K., Kubli, E., and Schäfer, M. A. (2006). Sex-peptide-regulated female sexual behavior requires a subset of ascending ventral nerve cord neurons. *Curr. Biol.* **16,** 1771–1782.

Spieth, H. T., and Ringo, J. M. (1983). Mating behavior and sexual isolation in *Drosophila*. In "The Genetics and Biology of *Drosophila*" (M. Ashburner, H. L. Carson, and J. N. Thompson, eds.), Vol. 3c, pp. 223–284. Academic Press, New York.

Swanson, W. J., and Vacquier, V. D. (1998). Concerted evolution in an egg receptor for a rapidly evolving abalone sperm protein. *Science* **281,** 710–712.

Swanson, W. J., and Vacquier, V. D. (2002). The rapid evolution of reproductive proteins. *Nat. Rev. Genet.* **3,** 137–144.

Swanson, W. J., Aquadro, C. F., and Vacquier, V. D. (2001). Polymorphism in abalone fertilization proteins is consistent with the neutral evolution of the egg's receptor for lysin (VERL) and positive darwinian selection of sperm lysin. *Mol. Biol. Evol.* **18,** 376–383.

Swanson, W. J., Wong, A., Wolfner, M. F., and Aquadro, C. F. (2004). Evolutionary expressed sequence tag analysis of *Drosophila* female reproductive tracts identifies genes subjected to positive selection. *Genetics* **168,** 1457–1465.

Takemori, N., and Yamamoto, M. T. (2009). Proteome mapping of the *Drosophila melanogaster* male reproductive system. *Proteomics* **9,** 2484–2493.

Tamura, K., Subramanian, S., and Kumar, S. (2004). Temporal patterns of fruit fly (*Drosophila*) evolution revealed by mutation clocks. *Mol. Biol. Evol.* **21,** 36–44.

Tompkins, L., Siegel, R. W., Gailey, D. A., and Hall, J. C. (1983). Conditioned courtship in *Drosophila* and its mediation by association of chemical cues. *Behav. Genet.* **13,** 565–578.

Tram, U., and Wolfner, M. F. (1998). Seminal fluid regulation of female sexual attractiveness in *Drosophila melanogaster. Proc. Natl. Acad. Sci. USA* **95,** 4051–4054.

Tsaur, S. C., and Wu, C. I. (1997). Positive selection and the molecular evolution of a gene of male reproduction, Acp26Aa of *Drosophila. Mol. Biol. Evol.* **14,** 544–549.

Tsaur, S. C., Ting, C. T., and Wu, C. I. (1998). Positive selection driving the evolution of a gene of male reproduction, Acp26Aa, of *Drosophila*: II Divergence versus polymorphism. *Mol. Biol. Evol.* **15,** 1040–1046.

Villella, A., and Hall, J. C. (2008). Neurogenetics of courtship and mating in *Drosophila. Adv. Genet.* **62,** 67–184.

Walker, M. J., Rylett, C. M., Keen, J. N., Audsley, N., Sajid, M., Shirras, A. D., and Isaac, R. E. (2006). Proteomic identification of *Drosophila melanogaster* male accessory gland proteins, including a pro-cathepsin and a soluble gamma-glutamyl transpeptidase. *Proteome Sci.* **4,** 9.

Watnick, T. J., Jim, Y., Matunis, E., Kernan, M. J., and Montell, C. (2003). A flagellar polycystin-2 homolog required for male fertility in *Drosophila. Curr. Biol.* **13,** 2179–2184.

Wedell, N., Gage, M. J. G., and Parker, G. A. (2002). Sperm competition, male prudence and sperm-limited females. *Trends Ecol. Evol.* **17,** 313–320.

Wigby, S., and Chapman, T. (2005). Sex peptide causes mating costs in female *Drosophila melanogaster*. *Curr. Biol.* **15,** 316–321.

Wigby, S., Sirot, L. K., Linklater, J. R., Buehner, N., Calboli, F. C. F., Bretman, A., Wolfner, M. F., and Chapman, T. (2009). Seminal fluid protein allocation and male reproductive success. *Curr. Biol.* **19,** 751–757.

Wolfner, M. F. (2009). Battle and ballet: Molecular interactions between the sexes in *Drosophila*. *J. Hered.* **100,** 399–410.

Wong, A., Albright, S. N., and Wolfner, M. F. (2006). Evidence for structural constraint on ovulin, a rapidly evolving *Drosophila melanogaster* seminal protein. *Proc. Natl. Acad. Sci. USA* **103,** 18644–18649.

Wong, A., Albright, S. N., Giebel, J. D., Ravi Ram, K., Ji, S. Q., Fiumera, A. C., and Wolfner, M. F. (2008a). A role for Acp29AB, a predicted seminal fluid lectin, in female sperm storage in *Drosophila melanogaster*. *Genetics* **180,** 921–931.

Wong, A., Turchin, M. C., Wolfner, M. F., and Aquadro, C. (2008b). Evidence for positive selection on *Drosophila melanogaster* seminal fluid protease homologs. *Mol. Biol. Evol.* **25,** 497–506.

Wong, A., Christopher, A. B., Buehner, N. A., and Wolfner, M. F. Immortal coils: Conserved dimerization motifs of the *Drosophila* ovulation prohormone ovulin. (Submitted for publication).

Xue, L., and Noll, M. (2000). *Drosophila* female sexual behavior induced by sterile males showing copulation complementation. *Proc. Natl. Acad. Sci. USA* **97,** 3272–3275.

Xue, L., and Noll, M. (2002). Dual role of the Pax gene paired in accessory gland development of *Drosophila*. *Development* **129,** 339–346.

Yamamoto, D. (2007). The neural and genetic substrates of sexual behavior in *Drosophila*. *Adv. Genet.* **59,** 39–66.

Yang, C. H., Belawat, P., Hafen, E., Jan, L. Y., and Jan, Y. N. (2008). *Drosophila* egg-laying site selection as a system to study simple decision-making processes. *Science* **319,** 1679–1683.

Yang, C. H., Rumpf, S., Xiang, Y., Gordon, M. D., Song, W., Jan, L. Y., and Jan, Y. N. (2009). Control of the postmating behavioral switch in *Drosophila* females by internal sensory neurons. *Neuron* **61,** 519–526.

Yapici, N., Kim, Y. J., Ribeiro, C., and Dickson, B. J. (2008). A receptor that mediates the postmating switch in *Drosophila* reproductive behaviour. *Nature* **451,** 33–37.

Yew, J. Y., Dreisewerd, K., Luftmann, H., Muthing, J., Pohlentz, G., and Kravitz, E. A. (2009). A new male sex pheromone and novel cuticular cues for chemical communication in *Drosophila*. *Curr. Biol.* **19,** 1245–1254.

3

Sleeping Together: Using Social Interactions to Understand the Role of Sleep in Plasticity

Jeffrey M. Donlea and Paul J. Shaw

Department of Anatomy and Neurobiology, Washington University School of Medicine, Campus Box 8108, St. Louis, Missouri, USA

ABSTRACT

Social experience alters the expression of genes related to synaptic function and plasticity, induces elaborations in the morphology of neural structures throughout the brain (Volkmar and Greenough, 1972; Greenough et al., 1978; Technau, 2007), improves cognitive and behavioral performance (Pham et al., 1999a; Toscano et al., 2006) and alters subsequent sleep (Ganguly-Fitzgerald et al., 2006). In this review,

Advances in Genetics, Vol. 68
Copyright 2009, Elsevier Inc. All rights reserved.
0065-2660/09 $35.00
DOI: 10.1016/S0065-2660(09)68003-2

we discuss the plastic mechanisms that are induced in response to social experience and how social enrichment can provide insight into the biological functions of sleep. © 2009, Elsevier Inc.

I. INTRODUCTION

Although research animals in laboratory environments are often housed individually in relatively simple enclosures, their wild counterparts must interact in more complex environments outside of a controlled setting. In nature, they must reliably find food, avoid enemies and predators, interact socially with conspecifics, and compete for potential mates. While standardized lab environments allow researchers to easily control environmental and social influences, manipulating the social environment of research animals can be a powerful experimental tool. Social experience alters the expression of genes related to synaptic function and plasticity, induces elaborations in the morphology of neural structures throughout the brain (Greenough *et al.*, 1978; Technau, 2007; Volkmar and Greenough, 1972), improves cognitive and behavioral performance (Pham *et al.*, 1999a; Toscano *et al.*, 2006), and alters subsequent sleep (Ganguly-Fitzgerald *et al.*, 2006). In this review, we discuss the use of social enrichment/isolation as an experimental paradigm to study plastic mechanisms in the brain and to investigate the relationship between sleep and synaptic plasticity.

Although specific details regarding the conditions of "enriched" environments may differ between studies (e.g., some enriched environments included novel objects that were periodically changed, some consisted of bare enclosures), all include enhanced social exposure. Thus, while nonsocial factors may contribute to some experimental results, it is likely that many of the experimental outcomes can be attributed to the enhanced social exposure. Greenough *et al.* (1978) compared the effect of two different social enrichment conditions on synaptic ultrastructure. In this study, the first enrichment condition consisted of a large cage in which 12 rats were housed and presented with toys that were changed daily as well as a 30-min opportunity to explore a different "toy-filled field." In the second enrichment condition, two rats were housed together in a standard laboratory cage. After 30 days in these conditions, rats were sacrificed and tissue from the occipital cortex was inspected for postsynaptic densities containing subsynaptic plate perforations, a marker of increased synaptic strength. While tissue from both enriched groups exhibited approximately 25% more subsynaptic plate perforations than tissue from isolated siblings, there was no difference detected between enriched conditions indicating that social interaction alone, not exposure to novel objects or a spatially larger enclosure, was likely sufficient to induce the changes in synaptic structure. Similarly, studies of environmental enrichment using the fruit fly *Drosophila melanogaster* have found significant changes in neural structure and behavior that are associated with social interactions and cannot be attributed to differences in enclosure volume or other physical differences in the housing conditions (Ganguly-Fitzgerald *et al.*, 2006; Heisenberg *et al.*, 1995).

II. SOCIAL ENRICHMENT INDUCES PLASTIC MECHANISMS

Following a period of enriched social experience, elaborations in neural circuitry have been characterized in a number of vertebrate and invertebrate species. Neurons in the visual cortex of socially enriched rats have an increased number of dendritic branches (Volkmar and Greenough, 1972) and show ultrastructural evidence of strengthened synaptic connections (Greenough *et al.*, 1978). Significant increases in the number of hippocampal synapses in socially enriched rats have also been reported (Briones *et al.*, 2006). Structural elaboration of individual cerebellar Purkinje cells in monkeys housed in a social environment suggests that the plastic effects of social enrichment on neural structures are evolutionarily conserved from rodents to primates (Floeter and Greenough, 1979). Elevated levels of neuronal growth factors may promote the increased dendritic elaboration and synaptogenesis that have been observed in response to social enrichment; exposure to an enriched social environment has been associated with increased expression of a number of neurotrophic factors including nerve growth factor (NGF) (Ickes *et al.*, 2000; Pham *et al.*, 1999a; Torasdotter *et al.*, 1998), brain-derived growth factor (BDNF) (Ickes *et al.*, 2000; Young *et al.*, 1999), and glial-cell-derived neurotrophic factor (GDNF) (Faherty *et al.*, 2005; Young *et al.*, 1999). Neurotrophic factors, particularly NGF and BDNF, play an important role in synaptic plasticity (Gómez-Palacio-Schjetnan and Escobar, 2008; Hennigan *et al.*, 2009) and impaired neurotrophin signaling results in memory impairments (Bekinschtein *et al.*, 2007; De Rosa *et al.*, 2005; Heldt *et al.*, 2007; Walz *et al.*, 2005).

The observations of structural plasticity in response to social enrichment seem to be widely conserved across species. Structural plasticity in response to enriched environments has been observed not only in mammals but also in several invertebrate species. Honeybees progress through a series of social roles within the hive structure that are associated with experience-dependent changes in the structure of brain regions including the mushroom bodies (MBs) (Withers *et al.*, 1993) and antennal lobes (Winnington *et al.*, 1996). Although the mechanisms that control the transition between social roles are not well characterized, these transitions are associated with extensive changes in gene expression profiles, indicating a role for genetic regulation (Whitfield *et al.*, 2003). While genetic studies are difficult to conduct using honeybees, the fruit fly

D. melanogaster is a widely used genetic model for the study of behavioral genetics. Indeed, social experience does induce significant changes in neuropil structure throughout the *Drosophila* brain (Heisenberg *et al.*, 1995; Technau, 2007). In these studies, the number of MB fibers were significantly elevated in socially enriched animals compared to wild-type siblings housed in isolation or deprived of visual or olfactory stimuli during social interactions. It should be noted that it is not yet clear whether the increased number of fibers results from increased branching of projections in existing neurons or the birth of new neurons. Interestingly, recent studies showed newly born Kenyon cells in the MBs of socially enriched wild-type flies (Technau, 2007). While little other evidence has been found for neurogenesis in the MBs of adult *Drosophila*, observations of newly born neurons have been made in the MBs of other insect species following social enrichment. BrdU labeling was used to reveal a significant elevation in the rate of neurogenesis in the MBs of adult crickets that were housed in a socially enriched environment for several days (Scotto-Lomassese *et al.*, 2000, 2002) and subsequent investigations found this elevation to be mediated via a nitric oxide-dependent signaling mechanism (Cayre *et al.*, 2005). Similar observations have been made in mice. Social enrichment resulted in an overall increase in the number of cells in the hippocampus including a significantly higher number of newly generated neurons and astroytes compared to socially isolated controls (Kempermann *et al.*, 1997). Based on their measurements, Kempermann *et al.* (1997) estimated that, on average, socially enriched mice possessed approximately 40,000 more dentate gyrus granule cells per hemisphere than their isolated siblings. These data suggest not only that the morphology of existing neurons can be altered by social experience but also that the generation of new neurons can be enhanced in response to the social enrichment.

As mentioned above, an increase in the number of MB fibers were found in socially enriched flies of a number of different wild-type strains. Interestingly, no change in fiber number was found in flies with mutant alleles of the classical memory genes *rutabaga* or *dunce* (Balling *et al.*, 2007). Together, these important studies suggest that complex sensory cues that accompany social interactions in the fly induce plastic mechanisms that alter the structure of the MBs, an important structure for associative memory in the fly. Further investigations have found that deprivation of visual stimuli by housing flies in complete darkness also alters the structure of several other neuropils involved in visual processing including the lamina, medulla, and lobula plate (Barth *et al.*, 1997) as well as the volume of the central complex and MB calyces (Barth and Heisenberg, 1997). Interestingly, Barth *et al.* (1997) describe a critical window for the effects of visual stimuli on the volume of the early visual system that is restricted to the first 4–5 days after eclosion; housing flies in complete darkness starting 5 days after eclosion had no effect on the volume of the lamina.

Conversely, another study showed that social experience can alter the volume of the MB calyces and medulla for at least 16 days after eclosion (Heisenberg *et al.*, 1995). Together, these findings suggest that experience-dependent plasticity may be controlled by two separate mechanisms. First, the early visual system (e.g., photoreceptor projections into the lamina) may be shaped by visual experience in the first few days after eclosion to optimize the ability of the brain to receive visual stimuli. Second, another type of plasticity that can encode memories throughout adult life can exhibit experience-dependent morphological changes for a much longer period of time and may allow downstream associative centers (including the MB) which are influenced by more complex, multisensory stimuli (including social interactions). This kind of temporal division can also be observed in mammalian models. While ocular dominance plasticity in the visual cortex can be robustly observed during a critical period early in life (reviewed in Berardi *et al.*, 2000), structural changes in other brain regions including the hippocampus can be observed much later in life (reviewed in Lee and Son, 2009).

III. FUNCTIONAL EFFECTS OF SOCIAL ENRICHMENT

Exposure to complex social environments not only alters neural circuitry at the structural level but also alters the physiological functioning of synapses in circuits throughout the brain. Hippocampal slices from rats housed in an enriched environment, for example, exhibit robust long-term potentiation (LTP) and long-term depression (LTD) in the Schaffer collateral pathway, while rats housed in social isolation demonstrate relative impairments for both LTP and LTD at this synapse (Artola *et al.*, 2006; Duffy *et al.*, 2001). Other studies, however, have found that hippocampal synapses in the medial perforant path can become significantly potentiated during social enrichment to the extent that LTP can be occluded (Foster *et al.*, 1996; Green and Greenough, 1986). Similar studies have found that mice housed in social isolation in lab conditions exhibit impaired LTP in the cingulate cortex when compared to tissue from wild mice that had developed in a natural, socially complex environment (Zhao *et al.*, 2009). Along with the physiological impairments observed in socially isolated animals, exposure to a socially enriched environment increases the expression of glutamatergic AMPA receptors throughout the hippocampus (Foster *et al.*, 1996), providing molecular evidence that glutamatergic synapses in the hippocampus can become highly potentiated during exposure to complex social environments. Because technical limitations have prevented the widespread use of electrophysiology to characterize the activity patterns of *Drosophila* central neurons until recently, very little is known about whether social enrichment induces similar physiological changes in the adult fly. Recent studies have, however, found that social enrichment significantly alters the excitability of motorneurons in larvae (Ueda and Wu, 2009).

Given the structural elaborations and physiological enhancements induced by socially enriched environments, it might be expected that these conditions improve the behavioral and cognitive performance of animals exposed to complex social interactions. Indeed, wild-type animals housed in an enriched environment demonstrate significant improvements in hippocampus-dependent memory in both the Morris water maze (Briones et al., 2006; Pham et al., 1999b; Toscano et al., 2006) and contextual fear conditioning (Duffy et al., 2001). Social enrichment seems to not only improve spatial memory in healthy animals but may also aid recovery from brain damage following traumatic brain injury, neonatal hypoxia-ischemia, excitotoxic injury, and early exposure to lead poisoning; in all of these conditions, enriched animals exhibited significant improvements in behavioral assays and normalized physiological functioning compared to socially isolated controls (Cao et al., 2008; Hoffman et al., 2008; Kline et al., 2007; Koh et al., 2005; Pereira et al., 2008, 2009; Young et al., 1999). Additionally, many of the cognitive benefits of social enrichment have also been observed in rodent models of neurodegenerative disease. In a rodent model of Parkinsonism, social enrichment during adulthood prevents the death of dopaminergic neurons in the substantia nigra (Faherty et al., 2005). Similar studies using a transgenic mouse model for Alzheimer's disease found that socially isolated mice expressing a mutant form of the human amyloid precursor protein (APP) exhibited more rapid decline in memory in the fear conditioning paradigm as well as a dramatic acceleration in amyloid-beta plaque deposition relative to socially housed animals that expressed the same mutant transgene (Dong et al., 2004). Importantly, recent studies of human populations have indicated that increased cognitive activity, including social interaction, throughout life may reduce the risk of developing Alzheimer's disease later in life (Carlson et al., 2008). Together, these data suggest that exposure to social enrichment may induce neuroprotective mechanisms to function normally in the face of genetic defects that would otherwise induce impairment. While the mechanisms that mediate these effects are not well characterized, the same neurotrophic signals that are elevated in response to social enrichment have been found to have neuroprotective effects in disease models for neurodegeneration; infusion of BDNF into the entorhinal cortex of aged mice expressing transgenic APP restores performance in the Morris water maze assay and restored expression of synaptic markers in the hippocampus (Nagahara et al., 2009). Similar effects have been observed by increasing NGF signaling (De Rosa et al., 2005) and there is preliminary evidence that these pathways may be effective therapeutic targets for humans afflicted with Alzheimer's disease (Tuszynski et al., 2005). Although the consequences of social interactions on

aging have not been well studied in *Drosophila*, recent experiments have suggested that social enrichment can delay premature death in flies mutant for Cu/
Zn superoxide dismutase (CuZnSOD) (Ruan and Wu, 2008), suggesting a phy-
logenetically conserved role for the functional benefits of social interactions. It is
important to note that lifespan extension was not observed in socially enriched
wild-type flies indicating that flies lacking CuZnSOD, which has been implicated
in mechanisms associated with a number of aging-related neurodegenerative
disorders including Parkinson's, Huntington's, and Alzheimer's diseases, may
be especially sensitive to beneficial effects of socially complex environments
and further implicates social enrichment as a potential intervention for the
alleviation of neurodegenerative disorders (Ruan and Wu, 2008).

IV. USING SOCIAL ENRICHMENT TO INVESTIGATE FUNCTIONS OF SLEEP

A. Sleep and plasticity

Although sleep is a biological process that is necessary for survival in vertebrates
and invertebrates, the underlying biological functions of sleep are currently
unknown. A growing body of literature, however, suggests an important and
evolutionarily conserved role for sleep in the processing and consolidation of
new memories. For example, hippocampal ensembles that were activated together when rats navigated a novel maze were reactivated in an identical manner
when the animals slept later (Wilson and McNaughton, 1994). Further studies
have found that similar reactivation can be observed in the visual cortex and that
replay in these two regions is coordinated to replay the same experience (Euston
et al., 2007; Ji and Wilson, 2007). There is also evidence for replay of waking
experience during sleep in humans. Subjects were recruited to play the video
game Tetris for several hours over the course of 3 days; over the course of the
study, a majority of the subjects reported hypnogogic imagery associated with
playing the game (Stickgold *et al.*, 2000). This type of imagery seemed to be
associated with the process of learning how to play the game because subjects
who had less previous Tetris experience were the most likely to report Tetris-
related imagery during dreams at night. This replay of newly formed associations
during sleep seems to facilitate the processing and consolidation of those memories. In one recent study, subjects were taught to play a card game while being
exposed to a specific odor cue. When subjects were reexposed to the same odor
cue during slow-wave sleep that night, hippocampal activation was significantly
elevated along with significantly improved hippocampus-dependent declarative
memory of the card game the next day (Rasch *et al.*, 2007). Additionally, human
performance in motor learning tasks stabilized with repeated trials over the

course of the day, then dramatically improved following a night of sleep (Stickgold and Walker, 2007; Walker *et al.*, 2005) and sleep deprivation impaired consolidation of long-term memory in rodents and flies (Ganguly-Fitzgerald *et al.*, 2006; Graves *et al.*, 2003). Indeed, sleep is significantly altered by previous experience and is important for the consolidation of recently acquired memories.

In the decade since the establishment of *Drosophila* as a model system for the study of sleep (Hendricks *et al.*, 2000; Shaw *et al.*, 2000), the fly has been successfully used to identify a number of candidate genes and pathways that are involved in sleep regulation. Among these pathways is the cAMP/PKA signaling cascade, which has classically been associated with learning and memory in the fly (Dudaí *et al.*, 1983). Several genetic mutants that induce a downregulation in cAMP signaling result in decreased sleep time and, conversely, manipulations that increase cAMP signaling yield elevated sleep compared to genetic controls (Hendricks *et al.*, 2001). Although the circuitry that controls sleep in the fly is largely unknown, additional investigations have found that disruption of the cAMP/PKA signaling cascade within the MBs, a structure that is important for associative processing, can strongly modulate sleep time (Joiner *et al.*, 2006).

Using a forward genetic screen to identify genes that alter sleep time, Cirelli *et al.* (2005) found that flies mutant for the voltage-dependent potassium channel *Shaker* (*Sh*) exhibit a robust decrease in sleep when compared to wild-type controls. Similarly, several mutants for the beta modulatory subunit *Hyperkinetic* (*Hk*), which influences *Sh* conductance, also sleep less than background controls (Bushey *et al.*, 2007). When these mutant flies were tested for memory using the heat-box assay (Putz and Heisenberg, 2002), flies with mutant alleles for *Sh* or *Hk* that resulted in decreased sleep time also exhibited memory impairments while alleles that did not change sleep also had no significant effect on memory performance (Bushey *et al.*, 2007). Although these results are purely correlational, they provide evidence that genetic mechanisms that influence sleep in the fly may be tightly intertwined with pathways that are important for learning and memory.

B. Social enrichment increases sleep

Despite the evidence for a relationship between plasticity and sleep, little is known about the mechanisms by which sleep and plasticity interact. As described above, social enrichment is a simple experimental manipulation that induces robust plasticity in circuits throughout the brain. Given that the social environment can induce structural changes in the brain in both mammals and invertebrates, it may be possible to begin to elucidate the underlying molecular mechanisms linking sleep and plasticity using the power of *Drosophila* genetics.

With that in mind, we evaluated flies that were housed in a socially enriched environment since they are likely to experience increased visual, olfactory phermonal, and auditory signals compared to siblings that are housed individually in social isolation. Moreover, interactions between individuals in a social environment are likely to increase the incidence of other behaviors including flying, jumping, geotaxis, and grooming, to name a few (Ganguly-Fitzgerald *et al.*, 2006). We hypothesized that increased exposure to these events would likely produce changes in structural plasticity within the brain which would increase sleep need. To test this hypothesis, wild-type flies were housed in a socially enriched environment with ∼35–40 siblings for 5 days. As seen in Fig. 3.1, socially enriched flies sleep approximately 2 h/day more than siblings that have been socially isolated for 5 days (Donlea *et al.*, 2009; Ganguly-Fitzgerald *et al.*, 2006) (Fig. 3.1). The increase is observed in both single-sex vials (male–male and female–female) and in mixed vials (male–female). Interestingly, social enrichment not only increases sleep time but, importantly, increases sleep consolidation as well. That is, isolated flies exhibit short-sleep bouts during the day which do not appear to be restorative (Seugnet *et al.*, 2008). In contrast, socially enriched flies maintain sleep bout durations more typical of that seen during night-time sleep (Ganguly-Fitzgerald *et al.*, 2006). Thus, social enrichment increases sleep consolidation sufficiently to permit the restorative properties of sleep.

Although the number of interactions that can potentially occur in the socially enriched environment is difficult to quantify precisely, it is clear that compared to their isolated siblings the brains of socially enriched flies must adapt and respond to a larger variety of stimuli. In that regard, it is worth noting that the change in sleep does not appear to be induced by other factors such as the volume of the enclosure or abiotic factors indirectly caused by enrichment. For example, blind and olfactory defective flies do not respond to social enrichment with an increase in sleep indicating that being in close proximity with 35–40 siblings for 5 days is not sufficient to induce a sufficient change in neuronal plasticity as to induce an increase in sleep (Ganguly-Fitzgerald *et al.*, 2006). Similarly, changes in sleep are not observed in enriched animals that are mutant for classical memory genes that also influence structural plasticity such as the adenylyl cyclase *rutabaga* or the cAMP phosphodiesterase *dunce*. Finally, the increase in sleep following social enrichment is directly proportional to the number of flies housed in the enriched environment (Ganguly-Fitzgerald *et al.*, 2006). Together, these observations indicate that the increase in sleep is likely due to social interactions inducing changes in neuronal plasticity.

An alternative hypothesis is that sleep is disrupted during social enrichment such that the subsequent increase in sleep simply reflects enhanced homeostatic drive not changes in plasticity. However, as mentioned above, flies mutant for memory genes do not exhibit an increase in sleep even though

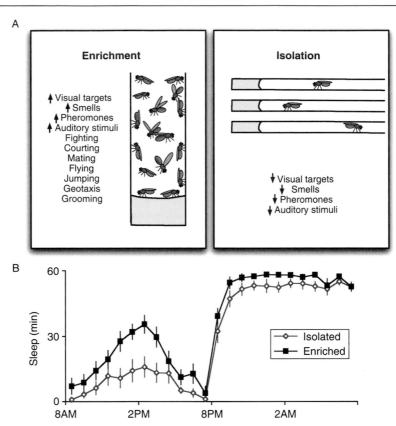

Figure 3.1. Exposure to an enriched social environment increases sleep in *Drosophila*. (A) Flies are
housed for 5 days either in a socially enriched vial with ∼40 other siblings or in small
vials in social isolation. During this time, flies in the socially enriched environment are
exposed to complex visual, olfactory, pheromonal, and auditory stimuli that are generated by other flies and fight, court, and mate with conspecific flies. Socially isolated
animals, however, are not exposed to these sensory cues and have no social interactions
with other flies. (B) When transferred to sleep monitors after 5 days of enrichment/
isolation, socially enriched females sleep ∼120 min/day more than their socially isolated
sisters. (See Page 2 in Color Section at the back of the book.)

one would expect that their sleep would be as disrupted during social enrichment
as that hypothesized to occur in wild-type flies. Interestingly, flies mutant for the
circadian clock gene *period* (per^{01}) exhibit a sleep rebound that is approximately
three times larger than wild-type flies, indicating that they are very responsive to
even small changes in sleep (Fig. 3.2). Despite their sensitivity to sleep loss,
however, per^{01} flies do not increase their sleep following social enrichment
(Fig. 3.2). Thus, if social enrichment resulted in disrupted sleep as the alternative

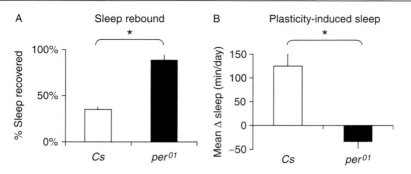

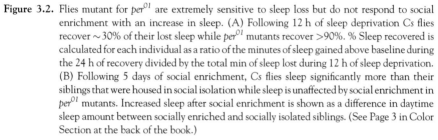

Figure 3.2. Flies mutant for per^{01} are extremely sensitive to sleep loss but do not respond to social enrichment with an increase in sleep. (A) Following 12 h of sleep deprivation Cs flies recover ~30% of their lost sleep while per^{01} mutants recover >90%. % Sleep recovered is calculated for each individual as a ratio of the minutes of sleep gained above baseline during the 24 h of recovery divided by the total min of sleep lost during 12 h of sleep deprivation. (B) Following 5 days of social enrichment, Cs flies sleep significantly more than their siblings that were housed in social isolation while sleep is unaffected by social enrichment in per^{01} mutants. Increased sleep after social enrichment is shown as a difference in daytime sleep amount between socially enriched and socially isolated siblings. (See Page 3 in Color Section at the back of the book.)

hypothesis suggests, per^{01} flies would show a larger increase in sleep than that seen in wild-type flies. Since per^{01} flies do not show an increase in sleep following social enrichment, we believe that the genetic mechanisms that control sleep homeostasis following sleep deprivation are dissociable from the mechanisms underlying increased sleep following social enrichment.

C. Circuits and genes that influence the response to social enrichment

To determine whether social enrichment may be dependent upon plastic mechanisms that are invoked following social interactions, we conducted a minibrain screen to restore *rutabaga* functioning in specific circuits in an otherwise *rutabaga* mutant background. We used the bipartite yeast-GAL4 system to express *rutabaga* in ~35 circuits (Brand and Perrimon, 1993). Surprisingly, when *rutabaga* was rescued in ventral lateral neurons (LN$_V$s), 16 cells that comprise the circadian clock, the increase in sleep following social enrichment was restored (Donlea et al., 2009). Recent studies from several independent groups have reported that the LN$_V$s play an important role in sleep–wake regulation in the fly (Agosto et al., 2008; Parisky et al., 2008; Sheeba et al., 2008). That is, manipulations that increase the excitability of the LN$_V$s result in large increases in waking behavior. A subset of the LN$_V$s innervates the medulla where they are uniquely situated to modulate the processing of simple and complex visual information. Together, these data suggest that environmentally induced plasticity in the LN$_V$s may be important for changes in sleep following social enrichment.

Having identified a circuit, the LN$_V$s, that is required for the social environment to induce an increase in sleep, we began to evaluate candidate genes that coordinate the changes in neuronal plasticity that are induced by social experience within this circuit. Although the clock gene *period* is widely expressed throughout the fly brain, specific rescue of *period* within clock cells was able to restore plasticity induced sleep in an otherwise *per^{01}* mutant background (Donlea et al., 2009). This observation is particularly interesting given that the *period* gene has been shown to play a noncircadian role in memory consolidation (Sakai et al., 2004). Similarly, the transcription factor *blistered* (*bs*) is both transcriptionally elevated in the brain following social enrichment and is required in the LN$_V$s for the social environment to modify sleep time. It is worth noting that the mammalian homolog of *blistered*, *serum response factor* induces transcription of a genetic program that mediates synaptic potentiation and is necessary both for *in vitro* assays of plasticity such as LTP and LTD as well as *in vivo* assays of behavioral plasticity (Etkin et al., 2006; Ramanan et al., 2005). To further explore the role of *bs* in coordinating social enrichment and sleep, we evaluated genes that are known to be regulated by *bs*. One such gene, *Epidermal growth factor receptor* (*Egfr*), is transcriptionally activated by social enrichment in wild-type flies. When *Egfr* was expressed in the LN$_V$s of a *bs* mutant, the increase in sleep following social enrichment was rescued. Thus, we have identified four genes that operate within a specific circuit that is known to both influence sleep–wake regulation and that is able to coordinate the interaction between social environment and sleep.

D. Social environment alters synaptic terminals

As mentioned above, Heisenberg and colleagues conducted a series of elegant studies demonstrating that it is possible to quantify environmentally induced structural changes in the brains of adult flies. To determine whether the social enrichment paradigm was able to induce structural changes in the brain, we expressed a GFP-tagged construct of the postsynaptic protein *discs-large* (UAS-*dlg*WT-gfp) in the LN$_V$s. As seen in Fig. 3.3, the number of synaptic terminals was significantly increased after 5 days of social enrichment. Similar results were obtained using a GFP tagged presynaptic marker. These results compliment those reported by Heisenberg and colleagues and demonstrate that the social environment is able to induce quantifiable changes in brain structures within a circuit that is known to play a role in sleep regulation. Since we have not demonstrated that the labeled synaptic terminals are functional, it is possible that the increased number of GFP-tagged terminals is an artifact of the environmental manipulation and does not reflect functional changes that influence sleep time. To test this possibility, UAS-*dlg*WT-gfp was expressed in a *bs* mutant background and flies were exposed to social enrichment. As seen in Fig. 3.3, the number of

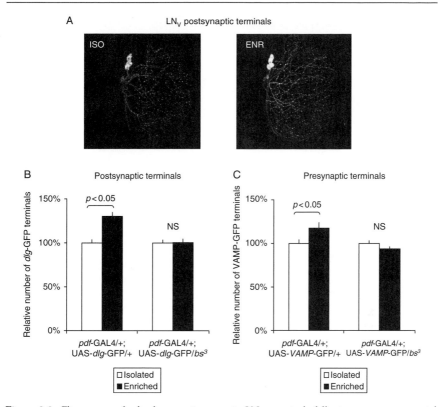

Figure 3.3. Flies mutant for *bs* show no increase in LN$_V$ terminals following exposure to social
enrichment. (A) LN$_V$ projections in socially isolated *pdf*-GAL4/+; UAS-*dlg*-GFP/+
flies (left, representative image) contain fewer GFP-positive terminals than those of
socially enriched siblings (right, representative image). (B) Following 5 days of social
enrichment, wild-type controls exhibit a significant increase in the number of LN$_V$
postsynaptic terminals labeled with *dlg*-GFP (left), but no change in the number of
postsynaptic terminals can be detected in flies carrying a mutant allele for *bs* (right).
(C) Similarly, an increased number of LN$_V$ presynaptic terminals labeled with *VAMP*-
GFP can be measured in wild-type control flies (left), but no change in presynaptic
terminal number was observed in flies with the *bs³* mutation (right). (See Page 3 in
Color Section at the back of the book.)

synaptic terminals was not altered in socially enriched *bs* mutants. That is, flies
that are not capable of responding to the social environment with an increase in
sleep do not display structural changes in the number of synaptic terminals.
These data suggest that in the absence of functional plasticity, the environment
is not associated with changes in structural plasticity.

E. Social enrichment versus long-term memory

Although structural plasticity and increased sleep following social enrichment require expression of several genes that are necessary for the formation and consolidation of associative memories, it is difficult to determine whether the increase in sleep is truly dependent upon plastic mechanisms. We hypothesized that if the increase in sleep following social enrichment is dependent upon plasticity-related processes then the circuits and genes that have been identified for social enrichment should also play a role in the consolidation of long-term memory. To test this hypothesis we utilized "courtship conditioning," an associative assay in which male flies learn to alter their courtship behavior based on previous exposure to unreceptive courtship targets. In courtship conditioning training, male flies are paired with mated female flies that are unreceptive to further copulation attempts or with male flies that have been genetically altered to express aphrodisiac pheromonal cues. During this training period, the subject male will proceed through stereotypical courtship behaviors in an attempt to woo the unreceptive courtship trainer, but is ultimately unable to copulate and forms an operant association between courtship rejection and normally aphrodisiac pheromonal cues (Gailey et al., 1984). Following training, male subject flies are returned to individual tubes. Subsequent memory is probed by exposing trained males to a normally attractive courtship target; if a trained male retains memory of the training experience, he will spend less time courting during the test period than his naïve brothers. When wild-type male flies are subjected to a spaced training protocol consisting of three 1-h training periods with a pheromonally feminized Tai2 male fly, they exhibit robust reductions in courtship for at least 48 h (Ganguly-Fitzgerald et al., 2006). Although acquisition and consolidation of memory following courtship conditioning requires similar genetic and neural mechanisms as memories formed in other associative assays (Siwicki and Ladewski, 2003), courtship conditioning requires male flies to process complex, naturalistic visual, and pheromonal cues and to interpret social behaviors and postures and, thus, seems more directly comparable to the types of neural processing that might occur during social enrichment than occurs in many other assays.

Following a spaced training protocol that induces long-term memories, male flies exhibit a significant increase in sleep. The increase in sleep appears necessary for memory consolidation since 4-h of sleep deprivation eliminates subsequent LTM. Interestingly, sleep deprivation immediately following training not only eliminates LTM but it also blocks the increase in sleep typically observed following the spaced-training protocol. This observation suggests that the increase in sleep following training is likely due to molecular processes associated with memory consolidation; once the memory has been disrupted there is no need for more sleep. Consistent with their role in experience-dependent changes in sleep, flies mutant for period and bs do not show increases in sleep following training and show no evidence for LTM after 48 h. Importantly, expression of either wild-type *per* or wild-type *bs* within the subset of clock neurons that are required for social enrichment, restores both the increase in sleep and LTM consolidation following courtship conditioning in *per* and *bs* mutant backgrounds, respectively. Thus, the genes and circuits that have been identified as playing a role in mediating the effects of social enrichment on sleep are also required for the increase in sleep following the formation of LTM. Together with the data demonstrating that social enrichment results in an increase in the number of synaptic terminals, it appears as if the social environment alters sleep time by modifying molecular processes associated with synaptic plasticity.

F. Synaptic homeostasis and sleep

Although the direct effects of sleep at the synapse are largely unknown, a recent hypothesis has suggested that a function of sleep may be to downscale synaptic connections throughout the brain (Tononi, 2003; Tononi and Cirelli, 2005). According to this hypothesis, experiences during waking drive patterns of activity in neural circuits that potentiate synapses and increase the strength and number of neural connections. If this kind of potentiation were allowed to continue unchecked, we could ultimately run into some severe consequences; energy requirements for the brain would grow, space available for new synaptic connections would disappear, and the signal strength between synaptic connections would reach a saturation point. To prevent these problems, Tononi and Cirelli have proposed that synchronized patterns of activity during sleep facilitate global synaptic downscaling to allow for normal functioning each morning (Tononi, 2003; Tononi and Cirelli, 2005). In support of this hypothesis, they have found that the levels of proteins that are associated with synaptic potentiation are increased during waking and decreased during sleep in both flies and rats (Gilestro *et al.*, 2009; Vyazovskiy *et al.*, 2008). Although these studies provide molecular data that are consistent with the downscaling hypothesis, they do not address whether sleep acts to reduce synaptic connections at the level of individual terminals.

Given the ability of the environment to increase synaptic terminals in the LN_Vs, social enrichment is uniquely suited to test the hypothesis that synaptic homeostasis is major role of sleep. If the synaptic homeostasis model is correct, then flies that were exposed to social enrichment for 5 days but are prevented from sleeping should show a persistence in the number of synaptic terminals. However, if consolidation is accomplished through synaptic potentiation then flies that are sleep deprived following social enrichment should display a decrease in the number terminals. Following social enrichment, flies were allowed to either sleep *ad libitum* for 48 h or were sleep deprived for 48 h. The number of synaptic terminals in projections from the large LN_Vs (l-LN_Vs) was then quantified. We found that the sleep-deprived flies retained an elevated number of l-LN_V

terminals following social enrichment, but those flies who could sleep *ad libidum* no longer had any significant change in terminal number when compared to their socially isolated siblings (Donlea *et al.*, 2009). These data are consistent with the downscaling hypothesis and indicate that flies sleep more when LN_V terminal number has been elevated by plastic mechanisms and that this increased sleep acts to reduce the number of terminals back to a baseline level.

At first glance, the increase in sleep following social enrichment appears somewhat paradoxical. That is, social enrichment increases the number of synaptic terminals in the LN_Vs while the activation of the LN_Vs strongly promotes waking (Sheeba *et al.*, 2008). How can these two observations be reconciled? During social enrichment, we hypothesize that complex sensory signals induce a prolonged elevation of LN_V activity (Fig. 3.4). The strength of these sensory signals results in an increase in the strength of synaptic connections between visual input circuits and the LN_Vs that is mediated by Hebbian mechanisms (Fig. 3.4B). The experience-dependent increase in synaptic strength seems to be implemented as an increase in the number of synaptic terminals in the LN_Vs over the course of social enrichment. After LN_V activity is elevated by several days of enriched social experience, it is possible that homeostatic mechanisms decrease the overall excitability of the LN_Vs to prevent chronic hyperexcitation and hold firing rate around a baseline set-point; similar homeostatic functioning has been previously described in *Drosophila* central synapses (Kazama and Wilson, 2008). Once enriched flies are moved into sleep monitors following social enrichment, complex sensory stimuli are removed and the lowered excitability of the LN_Vs reduces the firing rate below that of previously socially isolated controls (Fig. 3.4C). This lowered firing rate reduces the wake-promoting signal from these neurons and results in increased sleep for several days until LN_V excitability can be homeostatically elevated to restore a normalized firing rate. Coincidentally, this model could also account for the downscaling of synaptic terminal number following social enrichment; if overall excitability of the LN_Vs is reduced when flies are removed from social enrichment and complex sensory stimuli are removed, input to each individual terminal could become less likely to drive action potentials in the LN_Vs and, as a result, Hebbian mechanisms might weaken the synaptic connections by decreasing terminal numbers.

V. FUTURE DIRECTIONS

A. Local versus global regulation of experience-dependent sleep

Although sleep has been classically characterized as a unitary behavioral state that alters global physiology, processes that are associated with sleep are regulated on a local, circuit-dependent basis in the brain. Several species of marine

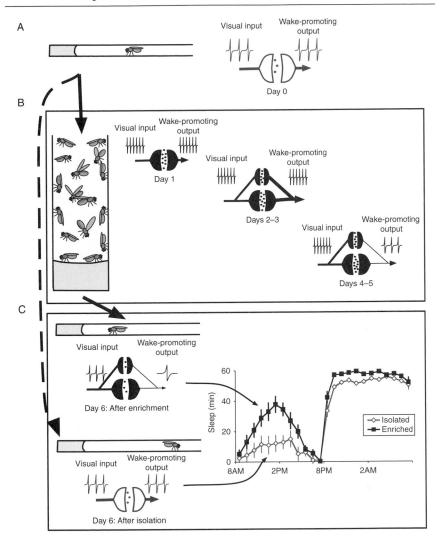

Figure 3.4. Proposed model for homeostatic regulation of synaptic terminals during social enrichment. (A) Under baseline conditions, wake-promoting output from the LN$_{V}$s is modulated by signals originating from visual circuits in the medulla. (B) When exposed to a socially enriched environment, complex sensory stimuli likely drives increased activity in primary visual circuits that provide input to the LN$_{V}$s. As a result, activity in the LN$_{V}$s would become more strongly correlated with input signals from sensory circuits and synaptic connections to the LN$_{V}$s become potentiated as new terminals are constructed. Following several days of hyperactivity and increased potentiation, homeostatic mechanisms may be induced in the LN$_{V}$s to reduce overall firing rate and prevent chronic hyperexcitation. (C) Upon transfer to sleep monitors, flies are withdrawn from

mammals, for instance, can exhibit electrophysiological patterns of sleep in one brain hemisphere while the other hemisphere appears to be awake (reviewed in Lyamin *et al.*, 2008). Recent studies have also found that sleep-associated patterns of activity can be significantly altered in local circuits by previous activity. That is, if a circuit is stimulated locally during waking, then the intensity of slow-wave activity (SWA) during subsequent sleep is elevated in that circuit (Cottone *et al.*, 2004; Huber *et al.*, 2004; Iwasaki *et al.*, 2004; Kattler *et al.*, 1994; Miyamoto *et al.*, 2003; Vyazovskiy *et al.*, 2000; Yasuda *et al.*, 2005). Conversely, if afferent activity to a cortical region is suppressed during the day, then SWA is locally reduced during sleep (Huber *et al.*, 2007). In all, these studies indicate that although sleep is a global behavioral state, specific circuits in the brain may "sleep" differently depending on their specific need and previous activity.

Our data suggest that plasticity in the LN_Vs can induce a robust increased sleep in response to social enrichment while other circuits that are known to play a role in learning and memory are much less effective (Donlea *et al.*, 2009). Interestingly, Gilestro *et al.* (2009) have reported that levels of the synaptic active zone protein *Bruchpilot* are dramatically elevated throughout much of the brain following 16 h of waking. These data suggest that synapses in a wide variety of circuits in the fly brain may become potentiated during waking. Indeed, rescue of *rutabaga* in all neurons results in the strongest increase in sleep following social enrichment consistent with the observation that waking influences many circuits besides just the LN_Vs. Thus, while the morphology of the LN_Vs and their projections make them well suited for quantifying synaptic terminals, the relationship between synaptic homeostasis and sleep will be enhanced through the identification of additional circuits with discrete expression patterns that can be modulated by experimental interventions. That is, while we favor the use of social environment to alter neuronal plasticity, other interventions may be useful for identifying additional circuits that can then be used to determine whether they are downscaled during sleep to the same degree as reported with this paradigm. Based on the vertebrate literature described above, it is likely that the degree of downscaling that is observed in each region will be proportional to the amount of local stimulation during waking. Further examination of synaptic homeostasis will be necessary to identify the mechanisms that mediate synaptic downscaling during sleep and to determine whether homeostatic downscaling might occur on a circuit-dependent basis during sleep.

---

the complex visual stimuli that are associated with a socially enriched environment. As a result, wake-promoting output from the LN_Vs is decreased and socially enriched flies sleep more than their siblings that had been housed in social isolation. (See Page 4 in Color Section at the back of the book.)

B. What social behaviors are happening during social enrichment?

Despite the use of *Drosophila* as a model system for behavioral genetics over the past several decades, relatively little is known about the naturalistic social behavior of the fly. Courtship and aggression behaviors that are used by male flies have been studied in detail and thoroughly reviewed elsewhere (Robin *et al.*, 2007; Villella and Hall, 2008), but social enrichment using only females induces a robust plastic response. The types of social interactions that occur in a female-only environment are not well characterized and the behavioral consequences of these plastic responses are entirely unknown. In order to better understand which neural circuits are likely to be altered by social enrichment as well as their functional roles, it would be helpful to better describe the types of social interactions that might occur during social enrichment. Two recent studies (Branson *et al.*, 2009; Dankert *et al.*, 2009) have utilized automated tracking algorithms to identify and score the social behavior of flies either in pairs or in a large group. The "ethomics" approach that these groups have begun to utilize might allow for rapid large-scale quantification of social behavior and to identify possible genetic and neural mechanisms that underlie these behaviors. As currently designed, the algorithms used in these two studies complement each other fairly well; the Ctrax software designed by Branson and colleagues is capable of tracking the movements of individual flies within larger groups over long periods of time while the CADABRA software suite utilized by Dankert and colleagues provides more detailed analysis of individual actions related to known courtship or aggression behaviors. These algorithms, however, are not optimally designed for the identification of novel types of interactions between individual animals; Ctrax does not provide detailed analysis of behavior and CADABRA is only capable of identifying predefined behavioral characteristics. Ultimately, these software tools may provide efficient tools for dissecting the mechanisms controlling known types of behavior, but only careful observation by human investigators is likely to allow for the identification and characterization of currently undescribed social interactions within groups of flies.

Although these interactions have not yet been characterized, it is likely that they are mediated, at least in part, through pheromonal communication. These chemical cues consist of a number of hydrocarbon compounds that are embedded in the waxy surface of the abdominal cuticle and are known to be crucial for the initiation and regulation of courtship behavior (Ferveur, 2005). Recent studies indicate that social context can significantly alter the composition of these hydrocarbon cues in male flies (Kent *et al.*, 2008; Krupp *et al.*, 2008). Interestingly, these studies found that altered pheromone production in response to social experience may have important consequences for future behavior; exposure to a genetically heterogeneous group induces wild-type males to alter the expression of courtship-related pheromonal cues and also results in increased mating frequency (Krupp *et al.*, 2008). Given the important role of chemosensation during enrichment in the induction of subsequent plasticity, it is possible that interactions via chemical cues comprise an important component of the social interactions that induce plastic changes in the brain. Further studies are needed to better characterize the neural circuits that are affected by pheromonal communication and the behavioral consequences of these signals on sleep. In conclusion, we expect that by examining how social interactions alter neuronal plasticity, we will identify novel roles of genes for sleep regulation as well as to better understand the role that these neural circuits play in modulating both daily sleep quotas and adaptive behavior.

References

Agosto, J., Choi, J. C., Parisky, K. M., Stilwell, G., Rosbash, M., and Griffith, L. C. (2008). Modulation of GABA(A) receptor desensitization uncouples sleep onset and maintenance in *Drosophila*. *Nat. Neurosci.* **11**, 354–359.

Artola, A., von Frijtag, J. C., Fermont, P. C., Gispen, W. H., Schrama, L. H., Kamal, A., and Spruijt, B. M. (2006). Long-lasting modulation of the induction of LTD and LTP in rat hippocampal CA1 by behavioural stress and environmental enrichment. *Eur. J. NeuroSci.* **23**, 261–272.

Balling, A., Technau, G. M., and Heisenberg, M. (2007). Are the structural changes in adult *Drosophila* mushroom bodies memory traces? Studies on biochemical learning mutants. *J. Neurogenet.* **21**, 209–217.

Barth, M., and Heisenberg, M. (1997). Vision affects mushroom bodies and central complex in *Drosophila melanogaster*. *Learn. Mem.* **4**, 219–229.

Barth, M., Hirsch, H. V., Meinertzhagen, I. A., and Heisenberg, M. (1997). Experience-dependent developmental plasticity in the optic lobe of *Drosophila melanogaster*. *J. Neurosci.* **17**, 1493–1504.

Bekinschtein, P., Cammarota, M., Igaz, L. M., Bevilaqua, L. R., Izquierdo, I., and Medina, J. H. (2007). Persistence of long-term memory storage requires a late protein synthesis- and BDNF-dependent phase in the hippocampus. *Neuron* **53**, 261–277.

Berardi, N., Pizzorusso, T., and Maffei, L. (2000). Critical periods during sensory development. *Curr. Opin. Neurobiol.* **10**, 138–145.

Brand, A. H., and Perrimon, N. (1993). Targeted gene expression as a means of altering cell fates and generating dominant phenotypes. *Development* **118**, 401–415.

Branson, K., Robie, A. A., Bender, J., Perona, P., and Dickinson, M. H. (2009). High-throughput ethomics in large groups of *Drosophila*. *Nat. Methods* **6**, 413–414.

Briones, T. L., Suh, E., Jozsa, L., and Woods, J. (2006). Behaviorally induced synaptogenesis and dendritic growth in the hippocampal region following transient global cerebral ischemia are accompanied by improvement in spatial learning. *Exp. Neurol.* **198**, 530–538.

Bushey, D., Huber, R., Tononi, G., and Cirelli, C. (2007). *Drosophila Hyperkinetic* mutants have reduced sleep and impaired memory. *J. Neurosci.* **27**, 5384–5393.

Cao, X., Huang, S., and Ruan, D. (2008). Enriched environment restores impaired hippocampal long-term potentiation and water maze performance induced by developmental lead exposure in rats. *Dev. Psychobiol.* **50**, 307–313.

Carlson, M. C., Helms, M. J., Steffens, D. C., Burke, J. R., Potter, G. G., and Plassman, B. L. (2008). Midlife activity predicts risk of dementia in older male twin pairs. *Alzheimer's Dementia: J. Alzheimer's Assoc.* **4**, 324–331.

Cayre, M., Malaterre, J., Scotto-Lomassese, S., Holstein, G. R., Martinelli, G. P., Forni, C., Nicolas, S., Aouane, A., Strambi, C., and Strambi, A. (2005). A role for nitric oxide in sensory-induced neurogenesis in an adult insect brain. Eur. J. NeuroSci. 21, 2893–2902.

Cirelli, C., Bushey, D., Hill, S. L., Huber, R., Kreber, R., Ganetzky, B., and Tononi, G. (2005). Reduced sleep in Drosophila Shaker mutants. Nature 434, 1087–1092.

Cottone, L. A., Adamo, D., and Squires, N. K. (2004). The effect of unilateral somatosensory stimulation on hemispheric asymmetries during slow wave sleep. Sleep 27, 63–68.

Dankert, H, Wang, L, Hoopfer, E, Anderson, DJ, and Perona, P (2009). Automated monitoring and analysis of social behavior in Drosophila. Nat. Methods 6, 297–303.

De Rosa, R., Garcia, A. A., Braschi, C., Capsoni, S., Maffei, L., Berardi, N., and Cattaneo, A. (2005). Intranasal administration of nerve growth factor (NGF) rescues recognition memory deficits in AD11 anti-NGF transgenic mice. Proc. Natl. Acad. Sci. USA 102, 3811–3816.

Dong, H., Goico, B., Martin, M., Csernansky, C. A., Bertchume, A., and Csernansky, J. G. (2004). Modulation of hippocampal cell proliferation, memory, and amyloid plaque deposition in APPsw (Tg2576) mutant mice by isolation stress. Neuroscience 127, 601–609.

Donlea, J. M., Ramanan, N., and Shaw, P. J. (2009). Use-dependent plasticity in clock neurons regulates sleep need in Drosophila. Science 324, 105–108.

Dudaí, Y., Uzzan, A., and Zvi, S. (1983). Abnormal activity of adenylate cyclase in the Drosophila memory mutant rutabaga. Neurosci. Lett. 42, 207–212.

Duffy, S. N., Craddock, K. J., Abel, T., and Nguyen, P. V. (2001). Environmental enrichment modifies the PKA-dependence of hippocampal LTP and improves hippocampus-dependent memory. Learn. Mem. 8, 26–34.

Etkin, A., Alarcón, J. M., Weisberg, S. P., Touzani, K., Huang, Y. Y., Nordheim, A., and Kandel, E. R. (2006). A role in learning for SRF: Deletion in the adult forebrain disrupts LTD and the formation of an immediate memory of a novel context. Neuron 50, 127–143.

Euston, D. R., Tatsuno, M., and McNaughton, B. L. (2007). Fast-forward playback of recent memory sequences in prefrontal cortex during sleep. Science 318, 1147–1150.

Faherty, C. J., Raviie Shepherd, K., Herasimtschuk, A., and Smeyne, R. J. (2005). Environmental enrichment in adulthood eliminates neuronal death in experimental Parkinsonism. Brain Res. Mol. Brain Res. 134, 170–179.

Ferveur, J. F. (2005). Cuticular hydrocarbons: Their evolution and roles in Drosophila pheromonal communication. Behav. Genet. 35, 279–295.

Floeter, M. K., and Greenough, W. T. (1979). Cerebellar plasticity: Modification of Purkinje cell structure by differential rearing in monkeys. Science 206, 227–229.

Foster, T. C., Gagne, J., and Massicotte, G. (1996). Mechanism of altered synaptic strength due to experience: Relation to long-term potentiation. Brain Res. 736, 243–250.

Gailey, D. A., Jackson, F. R., and Siegel, R. W. (1984). Conditioning mutations in Drosophila melanogaster affect an experience-dependent behavioral modification in courting males. Genetics 106, 613–623.

Ganguly-Fitzgerald, I., Donlea, J. M., and Shaw, P. J. (2006). Waking experience affects sleep need in Drosophila. Science 313, 1775–1781.

Gilestro, G. F., Tononi, G., and Cirelli, C. (2009). Widespread changes in synaptic markers as a function of sleep and wakefulness in Drosophila. Science 324, 109–112.

Gómez-Palacio-Schjetnan, A., and Escobar, M. L. (2008). In vivo BDNF modulation of adult functional and morphological synaptic plasticity at hippocampal mossy fibers. Neurosci. Lett. 445, 62–67.

Graves, L. A., Heller, E. A., Pack, A. I., and Abel, T. (2003). Sleep deprivation selectively impairs memory consolidation for contextual fear conditioning. Learn. Mem. 10, 168–176.

Green, E. J., and Greenough, W. T. (1986). Altered synaptic transmission in dentate gyrus of rats reared in complex environments: Evidence from hippocampal slices maintained in vitro. J. Neurophysiol. 55, 739–750.

Greenough, W. T., West, R. W., and DeVoogd, T. J. (1978). Subsynaptic plate perforations: Changes with age and experience in the rat. *Science* **202**, 1096–1098.

Heisenberg, M., Heusipp, M., and Wanke, C. (1995). Structural plasticity in the *Drosophila* brain. *J. Neurosci.* **15**, 1951–1960.

Heldt, S. A., Stanek, L., Chhatwal, J. P., and Ressler, K. J. (2007). Hippocampus-specific deletion of BDNF in adult mice impairs spatial memory and extinction of aversive memories. *Mol. Psychiatry* **12**, 656–670.

Hendricks, J. C., Finn, S. M., Panckeri, K. A., Chavkin, J., Williams, J. A., Sehgal, A., and Pack, A. I. (2000). Rest in *Drosophila* is a sleep-like state. *Neuron* **25**, 129–138.

Hendricks, J. C., Williams, J. A., Panckeri, K. A., Kirk, D., Tello, M. K., Yin, J. C., and Sehgal, A. (2001). A non-circadian role for cAMP signaling and CREB activity in *Drosophila* rest homeostasis. *Nat. Neurosci.* **4**, 1108–1115.

Hennigan, A., Callaghan, C. K., Kealy, J., Rouine, J., and Kelly, A. M. (2009). Deficits in LTP and recognition memory in the genetically hypertensive rat are associated with decreased expression of neurotrophic factors and their receptors in the dentate gyrus. *Behav. Brain Res.* **197**, 371–377.

Hoffman, A. N., Malena, R. R., Westergom, B. P., Luthra, P., Cheng, J. P., Aslam, H. A., Zafonte, R. D., and Kline, A. E. (2008). Environmental enrichment-mediated functional improvement after experimental traumatic brain injury is contingent on task-specific neurobehavioral experience. *Neurosci. Lett.* **431**, 226–230.

Huber, R., Ghilardi, M. F., Massimini, M., and Tononi, G. (2004). Local sleep and learning. *Nature* **430**, 78–81.

Huber, R., Esser, S. K., Ferrarelli, F., Massimini, M., Peterson, M. J., and Tononi, G. (2007). TMS-induced cortical potentiation during wakefulness locally increases slow wave activity during sleep. *PLoS ONE* **2**, e276.

Ickes, B. R., Pham, T. M., Sanders, L. A., Albeck, D. S., Mohammed, A. H., and Granholm, A. C. (2000). Long-term environmental enrichment leads to regional increases in neurotrophin levels in rat brain. *Exp. Neurol.* **164**, 45–52.

Iwasaki, N., Karashima, A., Tamakawa, Y., Katayama, N., and Nakao, M. (2004). Sleep EEG dynamics in rat barrel cortex associated with sensory deprivation. *NeuroReport* **15**, 2681–2684.

Ji, D., and Wilson, M. A. (2007). Coordinated memory replay in the visual cortex and hippocampus during sleep. *Nat. Neurosci.* **10**, 100–107.

Joiner, W. J., Crocker, A., White, B. H., and Sehgal, A. (2006). Sleep in *Drosophila* is regulated by adult mushroom bodies. *Nature* **441**, 757–760.

Kattler, H., Dijk, D. J., and Borbély, A. A. (1994). Effect of unilateral somatosensory stimulation prior to sleep on the sleep EEG in humans. *J. Sleep Res.* **3**, 159–164.

Kazama, H., and Wilson, R. I. (2008). Homeostatic matching and nonlinear amplification at identified central synapses. *Neuron* **58**, 401–413.

Kempermann, G., Kuhn, H. G., and Gage, F. H. (1997). More hippocampal neurons in adult mice living in an enriched environment. *Nature* **386**, 493–495.

Kent, C., Azanchi, R., Smith, B., Formosa, A., and Levine, J. D. (2008). Social context influences chemical communication in *D. melanogaster* males. *Curr. Biol.* **18**, 1384–1389.

Kline, A. E., Wagner, A. K., Westergom, B. P., Malena, R. R., Zafonte, R. D., Olsen, A. S., Sozda, C. N., Luthra, P., Panda, M., Cheng, J. P., and Aslam, H. A. (2007). Acute treatment with the 5-HT(1A) receptor agonist 8-OH-DPAT and chronic environmental enrichment confer neurobehavioral benefit after experimental brain trauma. *Behav. Brain Res.* **177**, 186–194.

Koh, S., Chung, H., Xia, H., Mahadevia, A., and Song, Y. (2005). Environmental enrichment reverses the impaired exploratory behavior and altered gene expression induced by early-life seizures. *J. Child. Neurol.* **20**, 796–802.

Krupp, J., Kent, C., Billeter, J., Azanchi, R., So, A., Schonfeld, J., Smith, B., Lucas, C., and Levine, J. D. (2008). Social experience modifies pheromone expression and mating behavior in male *Drosophila melanogaster*. *Curr. Biol.* **18,** 1373–1383.

Lee, E., and Son, H. (2009). Adult hippocampal neurogenesis and related neurotrophic factors. *BMB Rep.* **42,** 239–244.

Lyamin, O. I., Manger, P. R., Ridgway, S. H., Mukhametov, L. M., and Siegel, J. M. (2008). Cetacean sleep: An unusual form of mammalian sleep. *Neurosci. Biobehav. Rev.* **32,** 1451–1484.

Miyamoto, H., Katagiri, H., and Hensch, T. (2003). Experience-dependent slow-wave sleep development. *Nat. Neurosci.* **6,** 553–554.

Nagahara, A. H., Merrill, D. A., Coppola, G., Tsukada, S., Schroeder, B. E., Shaked, G. M., Wang, L., Blesch, A., Kim, A., Conner, J. M., Rockenstein, E., Chao, M. V., *et al.* (2009). Neuroprotective effects of brain-derived neurotrophic factor in rodent and primate models of Alzheimer's disease. *Nat. Med.* **15,** 331–337.

Parisky, K. M., Agosto, J., Pulver, S. R., Shang, Y., Kuklin, E., Hodge, J. J., Kang, K., Liu, X., Garrity, P. A., Rosbash, M., and Griffith, L. C. (2008). PDF cells are a GABA-responsive wake-promoting component of the *Drosophila* sleep circuit. *Neuron* **60,** 672–682.

Pereira, L. O., Strapasson, A. C., Nabinger, P. M., Achaval, M., and Netto, C. A. (2008). Early enriched housing results in partial recovery of memory deficits in female, but not in male, rats after neonatal hypoxia-ischemia. *Brain Res.* **1218,** 257–266.

Pereira, L. O., Nabinger, P. M., Strapasson, A. C., Nardin, P., Gonçalves, C. A., Siqueira, I. R., and Netto, C. A. (2009). Long-term effects of environmental stimulation following hypoxia-ischemia on the oxidative state and BDNF levels in rat hippocampus and frontal cortex. *Brain Res.* **1247,** 188–195.

Pham, T. M., Söderström, S., Winblad, B., and Mohammed, A. H. (1999a). Effects of environmental enrichment on cognitive function and hippocampal NGF in the non-handled rats. *Behav. Brain Res.* **103,** 63–70.

Pham, T. M., Ickes, B., Albeck, D., Söderström, S., Granholm, A. C., and Mohammed, A. H. (1999b). Changes in brain nerve growth factor levels and nerve growth factor receptors in rats exposed to environmental enrichment for one year. *Neuroscience* **94,** 279–286.

Putz, G., and Heisenberg, M. (2002). Memories in *Drosophila* heat-box learning. *Learn. Mem.* **9,** 349–359.

Ramanan, N., Shen, Y., Sarsfield, S., Lemberger, T., Schütz, G., Linden, D. J., and Ginty, D. D. (2005). SRF mediates activity-induced gene expression and synaptic plasticity but not neuronal viability. *Nat. Neurosci.* **8,** 759–767.

Rasch, B., Büchel, C., Gais, S., and Born, J. (2007). Odor cues during slow-wave sleep prompt declarative memory consolidation. *Science* **315,** 1426–1429.

Robin, C., Daborn, P. J., and Hoffmann, A. A. (2007). Fighting fly genes. *Trends Genet.* **23,** 51–54.

Ruan, H., and Wu, C. F. (2008). Social interaction-mediated lifespan extension of *Drosophila* Cu/Zn superoxide dismutase mutants. *Proc. Natl. Acad. Sci. USA* **105,** 7506–7510.

Sakai, T., Tamura, T., Kitamoto, T., and Kidokoro, Y. (2004). A clock gene, period, plays a key role in long-term memory formation in *Drosophila*. *Proc. Natl. Acad. Sci. USA* **101,** 16058–16063.

Scotto-Lomassese, S., Strambi, C., Strambi, A., Charpin, P., Augier, R., Aouane, A., and Cayre, M. (2000). Influence of environmental stimulation on neurogenesis in the adult insect brain. *J. Neurobiol.* **45,** 162–171.

Scotto-Lomassese, S., Strambi, C., Aouane, A., Strambi, A., and Cayre, M. (2002). Sensory inputs stimulate progenitor cell proliferation in an adult insect brain. *Curr. Biol.* **12,** 1001–1005.

Seugnet, L., Suzuki, Y., Vine, L., Gottschalk, L., and Shaw, P. J. (2008). D1 receptor activation in the mushroom bodies rescues sleep-loss-induced learning impairments in *Drosophila*. *Curr. Biol.* **18,** 1110–1117.

Shaw, P. J., Cirelli, C., Greenspan, R. J., and Tononi, G. (2000). Correlates of sleep and waking in *Drosophila melanogaster*. *Science* **287,** 1834–1837.

Sheeba, V., Fogle, K. J., Kaneko, M., Rashid, S., Chou, Y. T., Sharma, V. K., and Holmes, T. C. (2008). Large ventral lateral neurons modulate arousal and sleep in *Drosophila*. *Curr. Biol.* **18,** 1537–1545.

Siwicki, K. K., and Ladewski, L. (2003). Associative learning and memory in *Drosophila*: Beyond olfactory conditioning. *Behav. Processes* **64,** 225–238.

Stickgold, R. J., and Walker, M. P. (2007). Sleep-dependent memory consolidation and reconsolidation. *Sleep Med.* **8,** 331–343.

Stickgold, R. J., Malia, A., Maguire, D., Roddenberry, D., and O'Connor, M. (2000). Replaying the game: Hypnagogic images in normals and amnesics. *Science* **290,** 350–353.

Technau, G. M. (2007). Fiber number in the mushroom bodies of adult *Drosophila melanogaster* depends on age, sex and experience. *J. Neurogenet.* **21,** 183–196.

Tononi, G. (2003). Sleep and synaptic homeostasis: A hypothesis. *Brain Res. Bull.* **62,** 143–150.

Tononi, G., and Cirelli, C. (2005). Sleep function and synaptic homeostasis. *Sleep Med. Rev.* **10,** 49–62.

Torasdotter, M., Metsis, M., Henriksson, B. G., Winblad, B., and Mohammed, A. H. (1998). Environmental enrichment results in higher levels of nerve growth factor mRNA in the rat visual cortex and hippocampus. *Behav. Brain Res.* **93,** 83–90.

Toscano, C. D., McGlothan, J. L., and Guilarte, T. R. (2006). Experience-dependent regulation of zif268 gene expression and spatial learning. *Exp. Neurol.* **200,** 209–215.

Tuszynski, M. H., Thal, L., Pay, M., Salmon, D. P., Hoi Sang, U., Bakay, R., Patel, P., Blesch, A., Vahlsing, H. L., Ho, G., Tong, G., Potkin, S. G., *et al.* (2005). A phase 1 clinical trial of nerve growth factor gene therapy for Alzheimer disease. *Nat. Med.* **11,** 551–555.

Ueda, A, and Wu, CF (2009). Effects of social isolation on neuromuscular excitability and aggressive behaviors in *Drosophila*: Altered responses by *Hk* and *gsts1*, two mutations implicated in redox regulation. *J. Neurogenet.* Jun 30, 1–17. [Epub ahead of print].

Villella, A., and Hall, J. C. (2008). Neurogenetics of courtship and mating in *Drosophila*. *Adv. Genet.* **62,** 67–184.

Volkmar, F. R., and Greenough, W. T. (1972). Rearing complexity affects branching of dendrites in the visual cortex of the rat. *Science* **176,** 1445–1447.

Vyazovskiy, V., Borbély, A. A., and Tobler, I. (2000). Unilateral vibrissae stimulation during waking induces interhemispheric EEG asymmetry during subsequent sleep in the rat. *J. Sleep Res.* **9,** 367–371.

Vyazovskiy, V. V., Cirelli, C., Pfister-Genskow, M., Faraguna, U., and Tononi, G. (2008). Molecular and electrophysiological evidence for net synaptic potentiation in wake and depression in sleep. *Nat. Neurosci.* **11,** 200–208.

Walker, M. P., Stickgold, R., Alsop, D., Gaab, N., and Schlaug, G. (2005). Sleep-dependent motor memory plasticity in the human brain. *Neuroscience* **133,** 911–917.

Walz, R., Roesler, R., Reinke, A., Martins, M. R., Quevedo, J., and Izquierdo, I. (2005). Short- and long-term memory are differentialy modulated by hippocampal nerve growth factor and fibroblast growth factor. *Neurochem. Res.* **30,** 185–190.

Whitfield, C. W., Cziko, A. M., and Robinson, G. E. (2003). Gene expression profiles in the brain predict behavior in individual honey bees. *Science* **302,** 296–299.

Wilson, M. A., and McNaughton, B. L. (1994). Reactivation of hippocampal ensemble memories during sleep. *Science* **265,** 676–679.

Winnington, A. P., Napper, R. M., and Mercer, A. R. (1996). Structural plasticity of identified glomeruli in the antennal lobes of the adult worker honey bee. *J. Comp. Neurol.* **365,** 479–490.

Withers, G. S., Fahrbach, S. E., and Robinson, G. E. (1993). Selective neuroanatomical plasticity and division of labour in the honeybee. *Nature* **364,** 238–240.

Yasuda, T., Yasuda, K., Brown, R. A., and Krueger, J. M. (2005). State-dependent effects of light-dark cycle on somatosensory and visual cortex EEG in rats. *Am. J. Physiol. Regul. Integr. Comp. Physiol.* **289,** R1083–R1089.

Young, D., Lawlor, P. A., Leone, P., Dragunow, M., and During, M. J. (1999). Environmental enrichment inhibits spontaneous apoptosis, prevents seizures and is neuroprotective. *Nat. Med.* **5,** 448–453.

Zhao, M. G., Toyoda, H., Wang, Y. K., and Zhuo, M. (2009). Enhanced synaptic long-term potentiation in the anterior cingulate cortex of adult wild mice as compared with that in laboratory mice. *Mol. Brain* **2,** 11.

4

Approaching the Genomics of Risk-Taking Behavior

Alison M. Bell

School of Integrative Biology, University of Illinois, Urbana-Champaign, Urbana, Illinois 61801, USA

I. Introduction
II. Challenges and Approaches for Studying the Genetics of Risk-Taking Behavior in Humans and Other Animals
III. Approaching the Genomics of Risk-Taking Behaviors in Stickleback Fish
   A. Risk-taking behavior in sticklebacks
IV. Future Directions and Conclusions
   Acknowledgments
   References

ABSTRACT

Individual animals differ in their propensity to engage in dangerous situations, or in their risk-taking behavior. There is a heritable basis to some of this variation, but the environment plays an important role in shaping individuals' risk-taking propensity as well. This chapter describes some of the challenges in studying the genetic basis of individual differences in risk-taking behavior, arguing new insights will emerge from studies which take a whole-genome approach and which simultaneously consider both genetic and environmental influences on the behavior. The availability of genomic tools for three-spined stickleback, a small fish renowned for its variable behavior, opens up new possibilities for studying the genetic basis of natural, adaptive variation in risk-taking behavior. After introducing the general biology of sticklebacks, the chapter summarizes the

Advances in Genetics, Vol. 68
0065-2660/09 $35.00
DOI: 10.1016/S0065-2660(09)68004-4

existing literature on the genetic and environmental influences on risk-taking behavior, and describes the overall strategy that our group is taking to identify inherited and environmentally responsive genes related to risk-taking behavior in this species. Insights gleaned from such studies will be relevant to our understanding of similar behaviors in other organisms, including ourselves.   © 2009, Elsevier Inc.

I. INTRODUCTION

In a wide variety of organisms, individuals differ in their propensity to take risks in different situations. For example, mice show interindividual differences in behavior in an open field (Defries *et al.*, 1966), monkeys vary in aggressiveness (Suomi, 1987), and individual fish differ in their reaction to a predator (Huntingford, 1976). For the purposes of this review, I adopt a very inclusive definition of risk-taking behavior that can encompass this diversity. What is common to all of the examples above is that there is an element of danger involved in the behavior—from the chance of encountering a threat in a new environment, to the probability of injury in fights, to the possibility of death during interactions with a predator. Broadly construed, risk-taking behaviors are expressed in dangerous situations and increase the chance that an individual is injured or even dies.

Individual differences in the propensity to engage in risk-taking behavior have consequences for social groups for several reasons. First, risk-taking behaviors can influence other members of the social group. For example, high levels of aggression can influence other members of a social group by excluding nonaggressive individuals from access to resources, or via the effects of winning or losing fights on subsequent levels of aggression of others (the winner–loser effect). Diverse behaviors such as predator inspection behavior in fishes, mobbing behavior in birds, or alarm calling in rodents could reflect individual differences in the propensity to engage in dangerous situations around predators, and have consequences for the safety of other members of the group. Second, it is likely that the fitness of any given behavioral type (e.g., risk averse versus risk prone) depends on the frequency of other behavioral types within the social group, that is, fitness is frequency dependent. Therefore, understanding the causes and consequences of individual differences in risk-taking behavior would need to take into consideration the social context in which risk-taking behavior is expressed. Third, social dynamics could be influenced by the frequency of risk-taking phenotypes within the social group. Imagine a group composed entirely of risk-taking behavioral types, versus a group composed of risk-averse behavioral types. It is likely that such groups have different population dynamics and even fitness. Indeed, an intriguing question is whether there might be an optimal combination of behavioral types within the social group (Sih and Watters, 2005).

Finally, individual differences in exploratory behavior, or response to novelty, have consequences for dispersal and the colonization of new environments by a founding social group of individuals (Duckworth and Badyaev, 2007).

Individual differences in the propensity to engage in dangerous situations are often influenced by both genetic and environmental factors, and have relevance for human health and disease. Indeed, problems such as self-harm, addiction, sexual risk-taking, and violence have serious adverse consequences in humans. Despite the obvious costs to individuals and society and importance for health, we know relatively little about the etiology of risk-taking behaviors associated with these afflictions. While there are several candidate genes for related behaviors, candidate genes explain only a small fraction of the total genetic variation, suggesting that a whole-genome approach is likely to identify novel genes and pathways that are currently unknown. After describing some of the challenges that confront us as we try to understand the genetic basis of risk-taking behaviors in humans and other organisms, I describe the approach that we are taking to identify inherited and environmentally responsive genes related to risk-taking behaviors in a fresh new model system, three-spined stickleback fish.

II. CHALLENGES AND APPROACHES FOR STUDYING THE GENETICS OF RISK-TAKING BEHAVIOR IN HUMANS AND OTHER ANIMALS

There is a large literature investigating the genetic and neurochemical causes of variation in risk-taking behavior in humans and nonhuman animals. Some of the most recent findings have confirmed that human risk-taking behaviors related to psychopathology are influenced by inherited genetic factors (Kendler et al., 2003). Animal models for related behaviors such as *fearlessness* and *impulsiveness* (Strandberg et al., 2005), *fearfulness* (Boissy, 1995), *aggression* (Miczek et al., 2001), *anxiety and depression* (Adamec et al., 2006; Flint et al., 1995), and *substance dependence* (Crabbe, 2002) have extended these findings by elucidating the mechanisms underlying the specific behaviors. Progress in this field has been facilitated by the use of powerful tools such as microarrays (Kroes et al., 2006) and knockouts (Adamec et al., 2006) to find genes related to behavior.

However, many challenges remain as we try to study the genetic basis of risk-taking behavior, especially in identifying particular genes that might be related to variation in risk-taking behaviors. An immediate challenge is that like all complex traits, it is likely that a large number of genes, each of small effect, are involved in regulating risk-taking behavior (Kendler and Greenspan, 2006; Mackay, 2009). The sample sizes needed to identify quantitative trait loci (QTL) of small effect are prohibitively large ($n > 1000$), especially when we wish to understand how genes are responsive to the environment (Mackay, 2004; Plomin, 2005), and require huge numbers of markers (>1000) in order to find

QTL, making this an approach available to traditional model organisms, such as mouse (reviewed in Hovatta and Barlow, 2008; Willis-Owen and Flint, 2007) and some domesticated animals, for example, chickens (Wiren et al., 2009), quail (Beaumont et al., 2005), and cattle (Gutierrez-Gil et al., 2008).

Indeed, even when candidate genes have been proposed based on neurochemical hypotheses or QTL studies, specific candidate genes (e.g., DRD4, SERT, MAOA) for related behaviors only explain a small fraction of the total genetic variation (Reif and Lesch, 2003), indicating that we have yet to learn the identity of most of the important genes. Another recurring problem that has plagued candidate gene studies is the failure to replicate. While some studies, for example, find an association between the serotonin transporter polymorphism and behavior, other studies do not (reviewed in Reif and Lesch, 2003). At this point it is unclear how much of the inconsistency among studies is due to methodological problems, that is, population stratification, or due to real genetic differences between populations or species in the genetic mechanisms underlying similar behaviors.

Another challenge is that although there often appears to be some genetic component to risk-taking behavior, it cannot be denied that early experience affects related behaviors in both humans (Farrington, 2005) and nonhuman animals (Caldji et al., 2000; Meaney, 2001), and there is mounting evidence that the environment can influence behavior in a genotype-specific way (G × E interaction) (Caspi and Moffitt, 2006; Eaves et al., 2003). Arguably, the ubiquity (and effect size, >75% of the phenotypic variation in some cases) of genotype by environment interactions (Caspi and Moffitt, 2006; Kaufman et al., 2006) is an indication that studies will have the biggest impact if they simultaneously consider both genetic and environmental factors.

There has been considerable interest in recent years for using gene expression microarrays to identify genes related to behavior. Some of the advantages of measuring whole genome expression using microarrays are that the approach is unbiased and open-ended, and that it considers the coordinated action of the entire genome rather than focusing on one gene at a time. Moreover, microarrays are efficient, in that each microarray is its own self-contained experiment, which facilitates our ability to compare relative expression across genes. Also, microarrays are good for nonmodel systems in which it is not feasible to use traditional forward genetic approaches.

Behavioral experiments using microarrays have compared gene expression following different kinds of experiences, across different behavioral types, or among individuals from different populations or strains. Some studies have compared whole genome expression between groups that have or have not experienced some treatment or challenge that elicits a behavioral response, assuming that whatever genes are differentially expressed are related to the differing behavioral reactions. For example, studies have examined the genomic response to song in birds (London *et al.*, 2009), female response to courtship or mating in *Drosophila* (Carney, 2007; Kapelnikov *et al.*, 2008; Lawniczak and Begun, 2004; Mack *et al.*, 2006; McGraw *et al.*, 2008), swordtail fish (Cummings *et al.*, 2008), and honeybees (Kocher *et al.*, 2008), and male aggression in *Drosophila* (Wang *et al.*, 2008a,b). Related to risk-taking behavior, a number of studies have taken this strategy to identify gene expression correlates of anxiety in mice (Joo *et al.*, 2009; Sherrin *et al.*, 2009; Wang *et al.*, 2008a,b), *Drosophila* (Ibi *et al.*, 2008), monkeys (Sabatini *et al.*, 2007), and rats (Kabbaj *et al.*, 2004; Kroes *et al.*, 2006). However, many questions remain about how to interpret these experiments, how to compare across studies, and what treatment differences imply about the genetics of behavior (see Gibson, 2008). For example, it is difficult to know whether the expression differences are due to the application of the treatment rather than to the execution of the behavior itself. Moreover, the timing of sampling is critical; the genes involved in the immediate genomic response to the treatment could be different from those involved in the execution of the actual behavior. It is also unclear whether the same genetic mechanisms underlie treatment level differences and behavioral differences between individuals.

Another approach is to compare baseline gene expression differences between different genotypes or individuals that differ in behavior, assuming that whatever expression differences observed are related to behavioral differences (Aubin-Horth *et al.*, 2005; Ben-Shahar *et al.*, 2002; Dierick and Greenspan, 2006; Edwards *et al.*, 2006; Gammie *et al.*, 2007; Kim *et al.*, 2007, 2009; Quilter *et al.*, 2008; Renn *et al.*, 2008; Wang *et al.*, 2008a,b; Whitfield *et al.*, 2003). However, differences in transcript abundance could be a consequence, rather than a cause of differences in behavior. Despite this caveat, showing differential expression can be an important first step toward identifying the causative genetic polymorphisms underlying the difference in gene expression.

III. APPROACHING THE GENOMICS OF RISK-TAKING BEHAVIORS IN STICKLEBACK FISH

Obviously, we still have a lot to learn about the etiology of risk-taking behaviors in humans and other organisms. While traditional animal models have been useful in identifying mechanisms underlying some behaviors, such studies have typically been carried out on genetically homogeneous and lab-adapted strains. New insights might emerge from studying natural variation in behavior that resembles human behavioral variation.

Three-spined sticklebacks (*Gasterosteus aculeatus*) are renowned for their natural variation in behavior, morphology, and physiology (Bell and Foster, 1994). Therefore, in our work, we are using three-spined sticklebacks to identify candidate genes for natural variation in risk-taking behaviors. Sticklebacks are small (4–5 cm standard length at maturity) teleost fish that occur in the northern hemisphere. Sticklebacks commence breeding in the spring and typically live for 1 year. We are taking a whole-genome approach to understand genetic and environmental effects on suites of covarying risk-taking behaviors.

Our hypothesis is that there are inherited and environmentally responsive genes that affect risk-taking behaviors in sticklebacks, and those genes are shared with other animals, including humans. Later, we describe how risk-taking behaviors in this species show adaptive, natural variation within and between populations. Behaviors such as aggression and predator inspection are influenced by both inherited and experiential factors (Tulley and Huntingford, 1987), and they covary (Bell, 2005; Bell and Stamps, 2004; Huntingford, 1976). Moreover, the species is experimentally tractable—a favorite subject of the classical ethologists (Tinbergen, 1972; Wootton, 1984), the behaviors of sticklebacks are well characterized and are amenable to laboratory investigation. In addition, there are genomic resources for sticklebacks, including a whole-genome sequence (11× coverage by the Broad Institute), BAC and cDNA libraries, an EST project and a whole genome-linkage map (Peichel *et al.*, 2001).

Sticklebacks have another key attribute that makes them an especially good model: they have an unusual evolutionary history that has produced a replicated natural experiment. Freshwater populations of sticklebacks are the descendants of marine ancestors which independently colonized freshwater environments following glacial retreat ∼12,000 years ago. Isolated postglacial freshwater populations rapidly adapted to local environmental conditions, resulting in incredible phenotypic diversity among close relatives (Bell and Foster, 1994). Independent populations rapidly evolved convergent phenotypes in response to similar selective pressures, such that the same traits arose repeatedly and independently from the same ancestor. This system has already had great success in identifying the genetic basis of morphological traits (Colosimo *et al.*, 2005; Cresko *et al.*, 2004; Peichel *et al.*, 2001; Shapiro *et al.*, 2004).

This important feature of the stickleback system means that we have a built-in solution to a problem that has plagued candidate gene studies: the failure to replicate. After we identify a set of genes associated with risk-taking behaviors, we can replicate the experiment by comparing the genes in additional sets of populations. If the same genes are differentially expressed between an independent pair of risk-averse and risk-prone populations, we can conclude that the differential expression is related to the behavior and is not simply a consequence of other factors such as genetic drift (Fig. 4.1).

Figure 4.1. The replicated evolution of risk-taking behaviors in sticklebacks provides a natural experiment for testing candidate genes. This idealized map of Scotland shows the movement of sticklebacks from the ocean (on left) into freshwater rivers following glacial retreat. Sticklebacks inhabiting water bodies with abundant predators independently evolved increased levels of risk-taking behavior (dark fish) compared to sticklebacks inhabiting water bodies without predators (light fish).

A. Risk-taking behavior in sticklebacks

In the following sections, I describe how sticklebacks show natural variation in risk-taking behaviors such as predator inspection and aggression which are amenable to manipulative experimentation, and which resemble familiar human tendencies such as sensation-seeking, fearlessness, and disinhibition. Like the spectrum of externalizing behaviors (conduct disorder, antisocial personality, and substance abuse) in humans (Krueger *et al.*, 2002), risk-taking behaviors in sticklebacks also covary and are influenced by both heredity and early life stress. Another key similarity between the stickleback model of risk-taking and human externalizing behaviors is that they share the same neuroendocrine substrates, as described below. The recent availability of new genomic tools for sticklebacks means that these ecologically relevant behaviors are genetically tractable.

1. Individuals differ in their propensity to take risks in different situations

A consistent result of all of our studies is that there is substantial variation among both wild-caught and lab-reared individuals in how they behave in dangerous situations. For example, in an assay similar to the open field test (Yalcin *et al.*, 2004), some individual sticklebacks actively move around an unfamiliar and dangerous environment while others scarcely leave the safety of a refuge (Fig. 4.2). Therefore, we can use genetic correlates of this individual variation as a tool for finding candidate genes related to risk-taking behavior.

Another dangerous situation in which we observe individual differences in behavior is during confrontation by a potential predator. While some individuals hide in the presence of a predator, others swim up to the predator's mouth and face the predator head-on. The same individuals that engage in this dangerous behavior, known as predator inspection, are also relatively aggressive toward other sticklebacks and are more willing to take risks in order to gain food than their risk-averse conspecifics (Bell, 2005; Bell and Stamps, 2004; Huntingford, 1976).

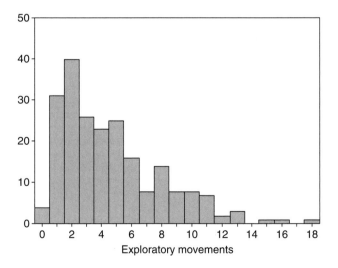

Figure 4.2. Individuals vary in their propensity to take risks. This histogram shows the distribution of exploratory movements of individuals in the presence of a predator. Methods: Juvenile sticklebacks from several populations were brought into the lab and the number of times an individual moved during 15 min in the presence of a predator was recorded. Mean = 4.6, S.D. = 3.39, n = 218.

Individual sticklebacks can be classified as either risk-prone or risk-averse in several other contexts. Predator inspection is one of the most obvious forms of risk-taking behaviors and has been widely studied in sticklebacks and other small fishes (e.g., Pitcher *et al.*, 1986). Despite the obvious danger involved in performing this behavior (Dugatkin and Godin, 1992; Milinski *et al.*, 1997), it is thought that predator inspection can provide reliable information about predation risk via both olfactory and visual cues.

Individual differences in risk-taking behavior are also observed when individuals balance the benefits of feeding against the costs of potential predation (Bell, 2005). Some risk-prone individuals are more willing to assume the risk of predation in order to get food than others, even when differences in size, sex, or hunger level are controlled for (e.g., Krause *et al.*, 1998). Therefore, differences in the willingness to forage in the presence of a predator can reflect intrinsic differences in the propensity to engage in risk-taking behaviors.

Fighting with conspecifics is another dangerous situation in which individual differences in risk-taking behavior are observed (Bell, 2005; Bell and Stamps, 2004; Huntingford, 1976). Intraspecific aggression in sticklebacks occurs during competition over access to resources, including food and territories and is manifested as biting, chasing, and attacking conspecifics. These behaviors are dangerous because in addition to energetic costs (Thorpe *et al.*, 1995), aggression can result in injury (Neat *et al.*, 1998) and exposure to predators while fighting (Diaz-Uriarte, 1999). The territorial aggression of male sticklebacks is especially well characterized (Bakker, 1994), but juveniles and females can also be aggressive (Bakker, 1986).

Interestingly, in sticklebacks, individuals that engage in higher levels of predator inspection are also more aggressive toward conspecifics (Bell, 2005; Bell and Stamps, 2004; Huntingford, 1976). Such covariation has parallels with the externalizing spectrum in humans, and with the observation that diverse psychiatric disorders in humans co-occur; antisocial behaviors, substance dependence, impulsivity, and behavioral disinhibition are comorbid (Kendler and Greenspan, 2006). Covariation among behavioral responses to dangerous situations in sticklebacks suggests that there is variation in the tendency to engage in risk-taking behaviors, and this tendency is manifested in multiple contexts (Sih *et al.*, 2004).

In humans, there is evidence that the broad, underlying tendency is more heritable than the particular manifestations (Krueger *et al.*, 2002). For example, whereas a significant portion of the variance in the externalizing latent trait can be traced to common genetic factors (>80% of the variance; Krueger *et al.*, 2002), heritabilities for single behaviors (e.g., conduct disorder, alcohol dependence) are generally much lower. If the same is true for risk-taking behaviors in sticklebacks, we might be more likely to find genes related to risk-taking behaviors if we look for genetic correlates of the propensity to be generally risk-averse or risk-prone.

2. There is a genetic basis to risk-taking behaviors in sticklebacks

Several lines of evidence suggest that there is a genetic component to risk-taking behaviors in sticklebacks. First, risk-taking behavior is repeatable. Figure 4.3 shows the results of an experiment in which different individual sticklebacks were measured for their behavioral reaction to a live fish predator (pike) on several occasions (Bell et al., 2009). Individuals differed in the total amount of time they spent freezing (remaining motionless) in the presence of a predator. Freezing is negatively related to risk-taking behaviors and we interpret it as reflecting fearfulness. Individual differences in freezing were consistent through time. Specifically, individuals that froze in response to the predator the first time they were measured also froze when measured on subsequent occasions. These data are important because they show that an individual's willingness to take risks is a stable attribute, supporting the hypothesis that there is a genetic basis to risk-taking behaviors in sticklebacks (Boake, 1994).

    Another line of evidence for a genetic component to risk-taking behavior is that there is significant genetic variation for risk-taking behavior among families (Bell, 2005; Bell and Stamps, 2004). When confronted with an unfamiliar environment, members of certain families actively explored the environment while members of other families did not. As can be seen from the data in Fig. 4.4,

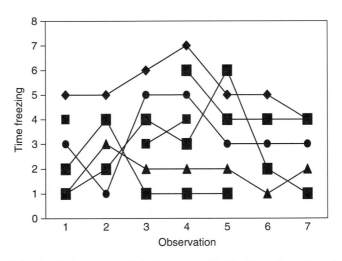

Figure 4.3. Risk-taking behavior in sticklebacks is repeatable. The Y-axis shows the rank order time spent freezing in the presence of a live predator. Time (the observation number) is on the X-axis. Each line represents a different individual; each point represents the measure of that individual on each observation day. Methods: On seven different occasions over the course of 2 months, individual sticklebacks were presented with a live pike and their behavioral response recorded. Repeatability was calculated as in Lessels and Boag (1987), $R = 0.68$, $F_{6,42} = 13$, $p < 0.001$. From Bell et al. (2009).

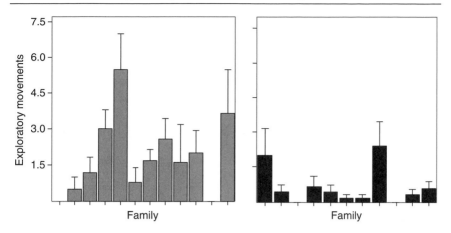

Figure 4.4. There is genetic variation for risk-taking behaviors among families. The data show the mean ($\pm$S.E.) number of exploratory movements in an unfamiliar environment of different families from two different populations, Putah Creek (gray) and the Navarro River (black). Methods: The exploratory behavior of lab-reared sticklebacks from 22 different families (2–5 full sibs/family) were measured at 7 months of age. Data were analyzed with a nested ANOVA with family and population as fixed factors (Population: $F_{21,102} = 21$, $p < 0.0001$; Family(Population): $F_{20,102} = 2.1$, $p = 0.003$).

there is substantial variation among families within populations in their willingness to explore an unfamiliar environment. These data are important because they show that there is heritable genetic variation within stickleback populations for risk-taking behaviors.

Common garden experiments also show that there is a genetic basis to risk-taking behaviors (Bell, 2005; Bell and Stamps, 2004). As can be seen from the data in Fig. 4.5, sticklebacks from one population were more willing to eat in the presence of a predator compared to sticklebacks from another population. This difference was apparent both for wild-caught parents and their lab-reared offspring that were reared under standardized laboratory conditions. These data are important because they show that the behavioral difference between the two populations has a genetic component. Moreover, it illustrates another important point about sticklebacks: the two populations differ in predation pressure which is a major source of variation for many stickleback phenotypes, and is discussed further later (Huntingford *et al.*, 1994; Reimchen, 1994).

A heritable basis to behaviors that deter predation (antipredator behaviors) is further substantiated by other studies on fishes, which have shown that differences between populations in antipredator behavior are inherited and arise without any experience of predators or without exposure to threatening situations (Magurran, 1990). There is also a heritable basis to aggression in sticklebacks (Bakker, 1986).

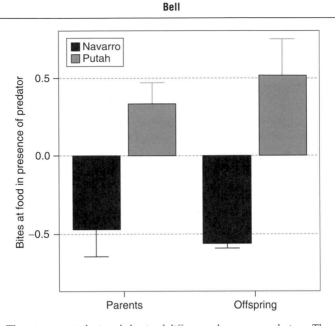

Figure 4.5. There is a genetic basis to behavioral differences between populations. These data show the standardized mean ($\pm$S.E.) number of bites at food while in the presence of a predator for wild-caught parents and their lab-reared offspring from two different populations, Navarro and Putah (Bell, 2005). Methods: The willingness to forage in the presence of a predator was measured on wild-caught adults from the two populations. Full-sib offspring from the two populations were reared under standardized laboratory conditions and measured for willingness to forage in the presence of a predator at adulthood. Both the parents and offspring of fish from Putah Creek took more bites in the presence of a predator (Parents: $F_{1,75} = 14.3$, $p < 0.0001$; Offspring: $F_{1,44} = 18.4$, $p < 0.0001$).

3. Risk-taking behavior is also influenced by the environment

However, we also know that risk-taking behaviors in sticklebacks are influenced by the environment, including early interactions with parents. During the reproductive season, which is triggered by lengthening day lengths and increasing temperatures, male sticklebacks defend nesting territories. Males attract females to spawn in their nests and defend the breeding territory from intruders and predators. After spawning, the female leaves the male's territory and the male is solely responsible for the care of the eggs. During the $\sim$9-day incubation period, the male "fans" (oxygenates) the eggs, removes rotten eggs and debris, and defends the territory. A breeding male stickleback tends his newly hatched offspring for $\sim$10 days. The fathers chase and catch fry that stray from the nest and spit them back into the nest. The fry are not injured during these interactions, but Tulley and Huntingford (1987), following a suggestion by Benzie

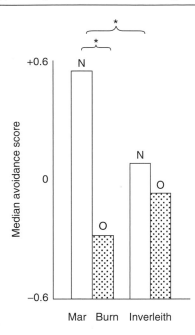

Figure 4.6. Paternal care improves antipredator behavior. This is Fig. 1 from Tulley and Huntingford (1987). The strength of avoidance shown to a model pike by sticklebacks from the Mar Burn, where predators are abundant, and from Inverleith Pond, where predators are absent with ("N") and without ("O") paternal care.

(1965), showed that these interactions help to prepare the stickleback fry for later encounters with predators. Figure 4.6, reproduced from Tulley and Huntingford (1987), shows that sticklebacks reared by their father ("N") avoided a model pike predator more than sticklebacks reared without paternal care (orphans, "O"). Interestingly, the behavioral difference between father-reared and orphan sticklebacks was only apparent among sticklebacks from a population where predators are abundant (the Mar Burn population on the left).

In addition to the influence of their father, young fishes also learn from their own direct experience with predators, and these early interactions affect subsequent risk-taking behaviors later in life (e.g., Vilhunen, 2006). Interestingly, the effect of both paternal care and direct experience appears to be especially strong for fish from high predation localities. The fact that fish from high predation localities learn faster about predators suggests an inherited predisposition to respond to experience (Huntingford and Wright, 1992). Similar G × E interactions, in which some genotypes (or populations) are more responsive to the environment, have been found for risk-taking behaviors in other fish species (e.g., Gerlai and Csanyi, 1990).

4. Sticklebacks from populations experiencing higher levels of predation engage in more risk-taking behaviors than their counterparts from safer environments

As mentioned above, sticklebacks are widely distributed throughout the northern hemisphere and have a penchant for rapidly adapting to their local environments. Populations that occur in similar environments have evolved similar behaviors. One of the most important selective forces shaping stickleback populations is predation pressure. That is, some lakes contain many predators which prey on sticklebacks ("high predation") while other lakes are relatively predator-free ("low predation"). We have been comparing the risk-taking behaviors of sticklebacks in a set of populations in Scotland that vary in predation pressure (Bell et al., 2009). Scotland is especially well suited for studying variation in risk-taking behavior in response to predation pressure because the country is teeming with postglacial waterbodies. Many lochs and ponds have been isolated for long enough (up to 15,000 generations) to independently evolve adaptations to high or low predation pressure and to become genetically differentiated from each other, but still capable of interbreeding (Malhi et al., 2006).

Interestingly, sticklebacks from areas where there are high levels of predation tend to be more risk-prone (i.e., they show higher levels of risk-taking behaviors) than their counterparts in safer environments (Huntingford and Coulter, 1989; Huntingford et al., 1994; Walling et al., 2003, 2004). This pattern has been documented in other small fish species as well, such as guppies (Magurran, 1986). While risky behavior in a dangerous environment might seem nonintuitive, this result is predicted by life history theory. The reason is that small individuals are especially vulnerable to predation, so when predation pressure is high, individuals that grow quickly will be favored because they are not small and vulnerable for long. Therefore, risk-taking behaviors that improve growth rate such as active foraging and aggression that results in access to resources should be favored when predation pressure is high (Mangel and Stamps, 2001). It is also worth noting that increased levels of risk-taking behaviors in humans have been documented in harsh or impoverished environments (Farrington, 2005; Kendler et al., 1995).

5. Risk-taking behaviors in sticklebacks resemble risk-taking behaviors in other organisms

Our contention that the genetic mechanisms underlying risk-taking behaviors in sticklebacks are shared with humans and other animals is supported by evidence that the neuroendocrine mechanisms underlying risk-taking behaviors are the same. If we can show that the neuroendocrine substrate underlying risk-taking behaviors in sticklebacks are conserved, then it is likely that the candidate genes that are identified in the stickleback system are likely to be good candidates for other animals, including humans.

For example, risk-taking behavior in sticklebacks is associated with serotonin turnover (Bell *et al.*, 2007). As can be seen in Fig. 4.7, individuals that engaged in more frequent predator inspection behavior had lower serotonergic activity. Low serotonergic activity is associated with sensation seeking in humans (Netter *et al.*, 1996) and aggression in several other animals (Higley *et al.*, 1996).

Second, breathing rate (opercular beats) is predictive of an individual's risk-taking tendency in sticklebacks. We measured opercular beats in response to handling stress and then the same individuals' behavioral reaction to a predator 24 h later. As can be seen from the data in Fig. 4.8, fish that breathed faster in response to handling stress were more likely to engage in risk-taking behavior (Bell *et al.*, 2009). Unlike some other physiological measures, opercular beats can

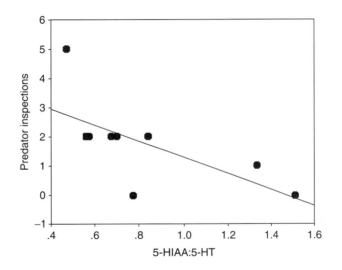

Figure 4.7. Risk-takers have lower serotoninergic activity. Predator inspection was negatively correlated with serotonergic activity, as measured by the turnover of serotonin (5-HT) to its metabolite 5-hydroxyphenylacetic acid (5-HIAA), 15 min after presentation of the predator. From Bell *et al.* (2007). Methods: Juvenile sticklebacks were measured for their behavioral reaction to a live pike predator and then sacrificed after 15 min, their brains removed and quickly deep frozen. Tissue was homogenized in 4% perchloric acid with an internal standard and monoamines were measured using HPLC with electrochemical detection as in Øverli *et al.* (1999). The correlation between individual levels of predator inspection and serotonergic activity was statistically significant ($r = -0.669$, $n = 9$, $p = 0.049$).

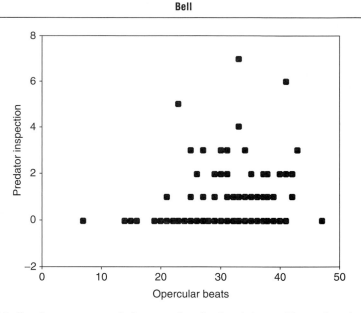

Figure 4.8. Breathing rate is an endophenotype for risk-taking behavior. The number of opercular
beats in 15 s after handling is positively correlated with predator inspection (Bell *et al.*,
2009). Methods: Individuals were subjected to brief handling stress and the number of
opercular beats was counted. At least 24 h after, the fish was presented with a live pike
and the number of predator inspections in 15 min was recorded. The partial correlation
coefficient (controlling for population) is statistically significant ($r=0.211$, $n=168$,
$p=0.006$) and was not related to body size.

be measured noninvasively and repeatedly on the same individuals. Therefore,
opercular beat rate is an appealing possible endophenotype linking genes and
behavior. Studies on other organisms have also found associations between
metabolic rate and risk-taking behaviors (Carere and Van Oers, 2004; Heim
and Nemeroff, 2001).

6. Genomic resources for sticklebacks

Genomic resources for stickleback research are rapidly expanding. The first
release of the stickleback genome, sequenced and assembled at the Broad
Institute, comprises 460 Mb. Gene models include approximately 21,000
known, novel and pseudogenes with an estimated 44,000 Genscan predictions.
Furthermore, close to 250,000, 3′ and 5′ expressed sequence tags (ESTs) from
various tissues and developmental stages have been submitted to GenBank and
comprise 15,087 Unigene clusters (Kingsley *et al.*, 2004).

IV. FUTURE DIRECTIONS AND CONCLUSIONS

Based on the work described earlier, we are taking a multipronged strategy for investigating the genetic correlates and causes of risk-taking behaviors in sticklebacks. We hypothesize that inherited and environmentally responsive genes that affect risk-taking behaviors in sticklebacks have orthologs with common function in other animal genomes, including human. Therefore, using whole genome microarrays, we are searching for candidate genes related to the propensity to engage in risk-taking behaviors that are both inherited and responsive to the environment. First, we are comparing baseline brain gene expression differences between risk-prone and risk-averse individuals and populations. Second, we are also identifying genes underlying risk-taking behaviors that are responsive to adverse environmental conditions by comparing the brain gene expression profiles of individuals exposed to different early life stressors. Third, we are testing candidate genes related to risk-taking behaviors in replicated populations of sticklebacks.

Future studies on the genetics of risk-taking behaviors in sticklebacks and other animals should focus on understanding the mechanisms by which candidate genes influence the behavior. Manipulations such as morpholinos, RNAi, and transgenics (Hosemann et al., 2004; Kingsley et al., 2004) will help to make the causal link between candidate genes and risk-taking behaviors; ultimately, making a risk-averse individual risk-prone. The future is also bright for whole genome association studies to study the genetic basis underlying phenotypes in natural populations, such as being done already in humans (Butcher et al., 2008; Dina et al., 2005; Donner et al., 2008; Doyle et al., 2008; Stallings et al., 2005) and dogs (Jones et al., 2008). In the not-too-distant future, it is likely that the main barrier to progress will not be genotyping our subjects, but in phenotyping the thousands of animals needed in order to obtain sufficient power to detect relationships between minor loci and behavior.

Acknowledgments

The author thanks Edelyn Verona for insights about anxiety, fear, and risk-taking behavior in humans and other organisms. This work is funded by NIH R01 GM082937 to A. M. B.

References

Adamec, R. E., Burton, P., et al. (2006). Vulnerability to mild predator stress in serotonin transporter knockout mice. Behav. Brain Res. **170,** 126–140.
Aubin-Horth, N., Landry, C. R., et al. (2005). Alternative life histories shape brain gene expression profiles in males of the same population. Proc. R. Soc. B-Biol. Sci. **272,** 1655–1662.
Bakker, T. C. M. (1986). Aggressiveness in sticklebacks (Gasterosteus aculeatus) a behavior-genetic study. Behaviour **98,** 1–144.

Bakker, T. C. M. (1994). Evolution of aggressive behaviour in the threespine stickleback. *In* "The Evolutionary Biology of the Threespine Stickleback" (M. A. Bell and S. A. Foster, eds.), pp. 345–379. Oxford University Press, Oxford.

Beaumont, C., Roussot, O., *et al.* (2005). A genome scan with AFLP((TM)) markers to detect fearfulness-related QTLs in Japanese quail. *Anim. Genet.* **36**, 401–407.

Bell, A. M. (2005). Differences between individuals and populations of threespined stickleback. *J. Evol. Biol.* **18**, 464–473.

Bell, M. A., and Foster, S. A. (1994). The Evolutionary Biology of the Threespine Stickleback. Oxford University Press, Oxford.

Bell, A. M., and Stamps, J. A. (2004). The development of behavioural differences between individuals and populations of stickleback. *Anim. Behav.* **68**, 1339–1348.

Bell, A. M., Backstrom, T., *et al.* (2007). Variable behavioral and neuroendocrine responses to ecologically-relevant challenges in sticklebacks. *Physiol. Behav.* **91**, 15–25.

Bell, A. M., Henderson, L., *et al.* (2009). Behavioral and respiratory responses to stressors in multiple populations of three-spined sticklebacks that differ in predation pressure. *J. Comp. Physiol.* (in press).

Ben-Shahar, Y., Robichon, A., *et al.* (2002). Influence of gene action across different time scales on behavior. *Science* **296**, 741–744.

Benzie, V. (1965). Some aspects of the anti-predator responses of two species of stickleback. PhD thesis *Zoology.* Oxford University.

Boake, C. R. B. (1994). Quantitative Genetic Studies of Behavioral Evolution. University of Chicago Press, Chicago.

Boissy, A. (1995). Fear and fearfulness in animals. *Quarterly Review of Biology* **70**, 165–191.

Butcher, L. M., Davis, O. S. P., *et al.* (2008). Genome-wide quantitative trait locus association scan of general cognitive ability using pooled DNA and 500 K single nucleotide polymorphism microarrays. *Genes Brain Behav.* **7**, 435–446.

Caldji, C., Diorio, J., *et al.* (2000). Variations in maternal care in infancy regulate the development of stress reactivity. *Biol. Psychiatry* **48**, 1164–1174.

Carere, C., and Van Oers, K. (2004). Shy and bold great tits (*Parus major*): Body temperature and breath rate in response to handling stress. *Physiol. Behav.* **82**, 905–912.

Carney, G. E. (2007). A rapid genome-wide response to *Drosophila melanogaster* social interactions. *BMC Genomics* **8.**

Caspi, A., and Moffitt, T. E. (2006). Opinion—Gene–environment interactions in psychiatry: Joining forces with neuroscience. *Nat. Rev. Neurosci.* **7**, 583–590.

Colosimo, P. F., Hosemann, K. E., *et al.* (2005). Widespread parallel evolution in sticklebacks by repeated fixation of ectodysplasin alleles. *Science* **307**, 1928–1933.

Crabbe, J. C. (2002). Genetic contributions to addiction. *Annu. Rev. Psychol.* **53**, 435–462.

Cresko, W. A., Amores, A., *et al.* (2004). Parallel genetic basis for repeated evolution of armor loss in Alaskan threespine stickleback populations. *Proc. Natl. Acad. Sci. USA* **101**, 6050–6055.

Cummings, M. E., Larkins-Ford, J., *et al.* (2008). Sexual and social stimuli elicit rapid and contrasting genomic responses. *Proc. R. Soc. B-Biol. Sci.* **275**, 393–402.

Defries, J. C., Hegmann, J. P., *et al.* (1966). Open-field behavior in mice—Evidence for a major gene effect mediated by visual system. *Science* **154**, 1577.

Diaz-Uriarte, R. (1999). Anti-predator behaviour changes following an aggressive encounter in the lizard *Tropidurus hispidus. Proc. R. Soc. Lond. B Biol. Sci.* **266**, 2457–2464.

Dierick, H. A., and Greenspan, R. J. (2006). Molecular analysis of flies selected for aggressive behavior. *Nat. Genet.* **38**, 1023–1031.

Dina, C., Nemanov, L., *et al.* (2005). Fine mapping of a region on chromosome 8p gives evidence for a QTL contributing to individual differences in an anxiety-related personality trait: TPQ harm avoidance. *Am. J. Med. Genet. Part B-Neuropsychiatr. Genet.* **132B**, 104–108.

Donner, J., Pirkola, S., et al. (2008). An association analysis of murine anxiety genes in humans implicates novel candidate genes for anxiety disorders. *Biol. Psychiatry* **64**, 672–680.

Doyle, A. E., Ferreira, M. A. R., et al. (2008). Multivariate genomewide linkage scan of neurocognitive traits and ADHD symptoms: Suggestive linkage to 3q13. *Am. J. Med. Genet. Part B-Neuropsychiatr. Genet.* **147B**, 1399–1411.

Duckworth, R. A., and Badyaev, A. V. (2007). Coupling of dispersal and aggression facilitates the rapid range expansion of a passerine bird. *Proc. Natl. Acad. Sci. USA* **104**, 15017–15022.

Dugatkin, L. A., and Godin J, J. (1992). Predator inspection shoaling and foraging under predation hazard in the Trinidadian guppy, *Poecilia reticulata. Environ. Biol. Fishes* **34**, 265–276.

Eaves, L. J., Silberg, J., et al. (2003). Resolving multiple epigenetic pathways to adolescent depression. *J. Child Psychol. Psychiatry* **44**, 1006–1014.

Edwards, A. C., Rollmann, S. M., et al. (2006). Quantitative genomics of aggressive behavior in *Drosophila melanogaster. PLoS Genet.* **2**, 1386–1395.

Farrington, D. P. (2005). Childhood origins of antisocial behavior. *Clin. Psychol. Psychother.* **12**, 177–190.

Flint, J., Corley, R., et al. (1995). A simple genetic basis for a complex psychological trait in laboratory mice. *Science* **269**, 1432–1435.

Gammie, S. C., Auger, A. P., et al. (2007). Altered gene expression in mice selected for high maternal aggression. *Genes Brain Behav.* **6**, 432–443.

Gerlai, R., and Csanyi, V. (1990). Genotype–environment interaction and the correlation structure of behavioral elements in paradise fish. *Physiol. Behav.* **47**, 343–356.

Gibson, G. (2008). The environmental contribution to gene expression profiles. *Nat. Rev. Genet.* **9**, 575–581.

Gutierrez-Gil, B., Ball, N., et al. (2008). Identification of quantitative trait loci affecting cattle temperament. *J. Hered.* **99**, 629–638.

Heim, C., and Nemeroff, C. B. (2001). The role of childhood trauma in the neurobiology of mood and anxiety disorders: Preclinical and clinical studies. *Biol. Psychiatry* **49**, 1023–1039.

Higley, J. D., Mehlman, P. T., et al. (1996). CSF testosterone and 5-HIAA correlate with different types of aggressive behaviors. *Biol. Psychiatry* **40**, 1067–1082.

Hosemann, K. E., Colosimo, P. E., et al. (2004). A simple and efficient microinjection protocol for making transgenic sticklebacks. *Behaviour* **141**, 1345–1355.

Hovatta, I., and Barlow, C. (2008). Molecular genetics of anxiety in mice and men. *Ann. Med.* **40**, 92–109.

Huntingford, F. A. (1976). The relationship between anti-predator behaviour and aggression among conspecifics in the three-spined stickleback. *Anim. Behav.* **24**, 245–260.

Huntingford, F. A., and Coulter, R. M. (1989). Habituation of predator inspection in the three-spined stickleback *Gasterosteus aculeatus. J. Fish Biol.* **35**, 153–154.

Huntingford, F. A., and Wright, P. J. (1992). Inherited population differences in avoidance conditioning in threespined sticklebacks, *Gasterosteus aculeatus. Behaviour* **122**, 264–273.

Huntingford, F. A., Wright, P. J., et al. (1994). Adaptive variation and antipredator behaviour in threespine stickleback. *In* "The Evolutionary Biology of the Threespine Stickleback" (M. A. Bell and S. A. Foster, eds.), pp. 277–295. Oxford University Press, Oxford.

Ibi, D., Takuma, K., et al. (2008). Social isolation rearing-induced impairment of the hippocampal neurogenesis is associated with deficits in spatial memory and emotion-related behaviors in juvenile mice. *J. Neurochem.* **105**, 921–932.

Jones, P., Chase, K., et al. (2008). Single-nucleotide-polymorphism-based association mapping of dog stereotypes. *Genetics* **179**, 1033–1044.

Joo, Y., Choi, K. M., *et al.* (2009). Chronic immobilization stress induces anxiety- and depression-like behaviors and decreases transthyretin in the mouse cortex. *Neurosci. Lett.* **461,** 121–125.

Kabbaj, M., Evans, S., *et al.* (2004). The search for the neurobiological basis of vulnerability to drug abuse: Using microarrays to investigate the role of stress and individual differences. *Neuropharmacology* **47,** 111–122.

Kapelnikov, A., Zelinger, E., *et al.* (2008). Mating induces an immune response and developmental switch in the *Drosophila* oviduct. *Proc. Natl. Acad. Sci. USA* **105,** 13912–13917.

Kaufman, J., Yang, B. Z., *et al.* (2006). Brain-derived neurotrophic factor-5-HTTLPR gene interactions and environmental modifiers of depression in children. *Biol. Psychiatry* **59,** 673–680.

Kendler, K. S., and Greenspan, R. J. (2006). The nature of genetic influences on behavior: Lessons from 'simpler' organisms. *Am. J. Psychiatry* **163,** 1683–1694.

Kendler, K. S., Kessler, R. C., *et al.* (1995). Stressful life events, genetic liability and onset of an episode of major depression in women. *Am. J. Psychiatry* **152,** 833–842.

Kendler, K. S., Jacobsen, K. C., *et al.* (2003). Specificity of genetic and environmental risk factors for use and abuse/dependence of cannabis, cocaine, hallucinogens, sedatives, stimulates and opiates in male twins. *Am. J. Psychiatry* **160,** 687–695.

Kim, S., Choi, K. H., *et al.* (2007). Suicide candidate genes associated with bipolar disorder and schizophrenia: An exploratory gene expression profiling analysis of post-mortem prefrontal cortex. *BMC Genomics* **8,** 413.

Kim, S., Zhang, S. M., *et al.* (2009). An E3 ubiquitin ligase, really interesting new gene (RING) finger 41, is a candidate gene for anxiety-like behavior and beta-carboline-induced seizures. *Biol. Psychiatry* **65,** 425–431.

Kingsley, D. M., Zhu, B. L., *et al.* (2004). New genomic tools for molecular studies of evolutionary change in threespine sticklebacks. *Behaviour* **141,** 1331–1344.

Kocher, S. D., Richard, F. J., *et al.* (2008). Genomic analysis of post-mating changes in the honey bee queen (*Apis mellifera*). *BMC Genomics* **9,** 232.

Krause, J., Loader, S. P., *et al.* (1998). Refuge use by fish as a function of body length-related metabolic expenditure and predation risks. *Proc. R. Soc. Lond. Ser. B* **265,** 2373–2379.

Kroes, R. A., Panksepp, J., *et al.* (2006). Modeling depression: Social dominance–submission gene expression patterns in rat neocortex. *Neuroscience* **137,** 37–49.

Krueger, R. F., Hicks, B. M., *et al.* (2002). Etiologic connections among substance dependence, antisocial behavior, and personality: Modeling the externalizing spectrum. *J. Abnorm. Psychol.* **111,** 411–424.

Lawniczak, M. K. N., and Begun, D. J. (2004). A genome-wide analysis of courting and mating responses in *Drosophila melanogaster* females. *Genome* **47,** 900–910.

Lessels, C. M., and Boag, P. T. (1987). Unrepeatable repeatabilities: A common mistake. *The Auk* **104,** 116–121.

London, S. E., Dong, S., *et al.* (2009). Developmental shifts in gene expression in the auditory forebrain during the sensitive period for song learning. *Dev. Neurobiol.* **69,** 437–450.

Mack, P. D., Kapelnikov, A., *et al.* (2006). Mating-responsive genes in reproductive tissues of female *Drosophila melanogaster*. *Proc. Natl. Acad. Sci. USA* **103,** 10358–10363.

Mackay, T. F. C. (2004). The genetic architecture of quantitative traits: Lessons from *Drosophila*. *Curr. Opin. Genet. Dev.* **14,** 253–257.

Mackay, T. F. C. (2009). The genetic architecture of complex behaviors: Lessons from *Drosophila*. *Genetica* **136,** 295–302.

Magurran, A. E. (1986). Predator inspection behaviour in minnow shoals: Differences between populations and individuals. *Behav. Ecol. Sociobiol.* **19,** 267–273.

Magurran, A. E. (1990). The inheritance and development of minnow anti-predator behaviour. *Anim. Behav.* **39,** 834–842.

Malhi, R. S., Rhett, G., *et al.* (2006). Mitochondrial DNA evidence of an early holocene population expansion of threespine sticklebacks from Scotland. *Mol. Phylogenet. Evol.* **40,** 148–154.

Mangel, M., and Stamps, J. (2001). Trade-offs between growth and mortality and the maintenance of individual variation in growth. *Evol. Ecol. Res.* **3,** 583–593.

McGraw, L. A., Clark, A. G., *et al.* (2008). Post-mating gene expression profiles of female *Drosophila melanogaster* in response to time and to four male accessory gland proteins. *Genetics* **179,** 1395–1408.

Meaney, M. J. (2001). Maternal care, gene expression, and the transmission of individual differences in stress reactivity across generations. *Annu. Rev. Neurosci.* **24,** 1161–1192.

Miczek, K. A., Maxson, S. C., *et al.* (2001). Aggressive behavioral phenotypes in mice. *Behav. Brain Res.* **125,** 167–181.

Milinski, M., Luthi, J. H., *et al.* (1997). Cooperation under predation risk: Experiments on costs and benefits. *Proc. R. Soc. Lond. B Biol. Sci.* **264,** 831–837.

Neat, F. C., Taylor, A. C., *et al.* (1998). Proximate costs of fighting in male cichlid fish: The role of injuries and energy metabolism. *Anim. Behav.* **55,** 875–882.

Netter, P., Henning, J., *et al.* (1996). Serotonin and dopamine as mediators of sensation seeking behavior. *Neuropsychobiology* **34,** 155–165.

Øverli, O., Harris, C. A., *et al.* (1999). Short-term effects of fights for social dominance and the establishment of dominant–subordinate relationships on brain monoamines and cortisol in rainbow trout. *Brain Behav. Evol.* **54,** 263–275.

Peichel, C. L., Nereng, K. S., *et al.* (2001). The genetic architecture of divergence between threespine stickleback species. *Nature* **414,** 901–905.

Pitcher, T. J., Green, D. A., *et al.* (1986). Dicing with death: Predator inspection behavior in minnow (*Phoxinus phoxinus*) shoals. *J. Fish Biol.* **28,** 439–448.

Plomin, R. (2005). Finding genes in child psychology and psychiatry: When are we going to be there? *J. Child Psychol. Psychiatry* **46,** 1030–1038.

Quilter, C. R., Gilbert, C. L., *et al.* (2008). Gene expression profiling in porcine maternal infanticide: A model for puerperal psychosis. *Am. J. Med. Genet. Part B-Neuropsychiatr. Genet.* **147B,** 1126–1137.

Reif, A., and Lesch, K. P. (2003). Toward a molecular architecture of personality. *Behav. Brain Res.* **139,** 1–20.

Reimchen, T. E. (1994). Predators and morphological evolution in threespine stickleback. *In* "The Evolutionary Biology of the Threespine Stickleback" (M. A. Bell and S. A. Foster, eds.), pp. 240–273. Oxford University Press, Oxford.

Renn, S. C. P., Aubin-Horth, N., *et al.* (2008). Fish and chips: Functional genomics of social plasticity in an African cichlid fish. *J. Exp. Biol.* **211,** 3041–3056.

Sabatini, M. J., Ebert, P., *et al.* (2007). Amygdala gene expression correlates of social behavior in monkeys experiencing maternal separation. *J. Neurosci.* **27,** 3295–3304.

Shapiro, M. D., Marks, Peichel, C. L., *et al.* (2004). Genetic and developmental basis of evolutionary pelvic reduction in threespine sticklebacks. *Nature* **428,** 717–723.

Sherrin, T., Blank, T., *et al.* (2009). Region specific gene expression profile in mouse brain after chronic corticotropin releasing factor receptor 1 activation: The novel role for diazepam binding inhibitor in contextual fear conditioning. *Neuroscience* **162,** 14–22.

Sih, A., and Watters, J. V. (2005). The mix matters: Behavioural types and group dynamics in water striders. *Behaviour* **142,** 1417–1431.

Sih, A., Bell, A. M., *et al.* (2004). Behavioral syndromes: An integrative overview. *Q. Rev. Biol.* **79,** 241–277.

Stallings, M. C., Corley, R. P., *et al.* (2005). A genome-wide search for quantitative trait loci that influence antisocial drug dependence in adolescence. *Arch. Gen. Psychiatry* **62,** 1042–1051.

Strandberg, E., Jacobsson, J., *et al.* (2005). Direct genetic, maternal and litter effects on behaviour in German shepherd dogs in Sweden. *Livestock Prod. Sci.* **93,** 33–42.

Suomi, J. S. (1987). Genetic and maternal contributions to individual differences in Rhesus monkey biobehavioral development. *In* "Psychobiological Aspects of Behavioral Development" (N. Krasnagor, ed.), pp. 397–419. Academic Press, New York.

Thorpe, K. E., Taylor, A. C., *et al.* (1995). How costly is fighting? Physiological effects of sustained exercise and fighting in swimming crabs, *Necora puber. Anim. Behav.* **50,** 1657–1666.

Tinbergen, N. (1972). The Animal in its World; Explorations of an Ethologist. Harvard University Press, Cambridge.

Tulley, J. J., and Huntingford, F. A. (1987). Paternal care and the development of adaptive variation in anti-predator responses in sticklebacks. *Anim. Behav.* **35,** 1570–1572.

Vilhunen, S. (2006). Repeated antipredator conditioning: A pathway to habituation or to better avoidance? *J. Fish Biol.* **68,** 25–43.

Walling, C. A., Dawnay, N., *et al.* (2003). Do competing males cooperate? Familiarity and its effect on cooperation during predator inspection in male three-spined sticklebacks (*Gasterosteus aculeatus*). *J. Fish Biol.* **63,** 243–244.

Walling, C. A., Dawnay, N., *et al.* (2004). Predator inspection behaviour in three-spined sticklebacks (*Gasterosteus aculeatus*): Body size, local predation pressure and cooperation. *Behav. Ecol. Sociobiol.* **56,** 164–170.

Wang, J., Ross, K. G., *et al.* (2008a). Genome-wide expression patterns and the genetic architecture of a fundamental social trait. *PLoS Genet.* **4,** e1000127.

Wang, L. M., Dankert, H., *et al.* (2008b). A common genetic target for environmental and heritable influences on aggressiveness in *Drosophila. Proc. Natl. Acad. Sci. USA* **105,** 5657–5663.

Whitfield, C. W., Cziko, A. M., *et al.* (2003). Gene expression profiles in the brain predict behavior in individual honey bees. *Science* **302,** 296–299.

Willis-Owen, S. A. G., and Flint, J. (2007). Identifying the genetic determinants of emotionality in humans; insights from rodents. *Neurosci. Biobehav. Rev.* **31,** 115–124.

Wiren, A., Gunnarsson, U., *et al.* (2009). Domestication-related genetic effects on social behavior in chickens—Effects of genotype at a major growth quantitative trait locus. *Poult. Sci.* **88,** 1162–1166.

Wootton, R. J. (1984). A Functional Biology of Sticklebacks. University of California Press, Berkeley.

Yalcin, B., Willis-Owen, S. A. G., *et al.* (2004). Genetic dissection of a behavioral quantitative trait locus shows that Rgs2 modulates anxiety in mice. *Nat. Genet.* **36,** 1197–1202.

Index

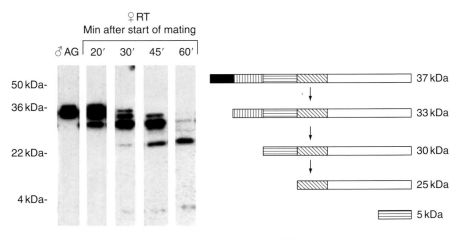

Chapter 2, Figure 2.3 (See Page 36 of this volume).

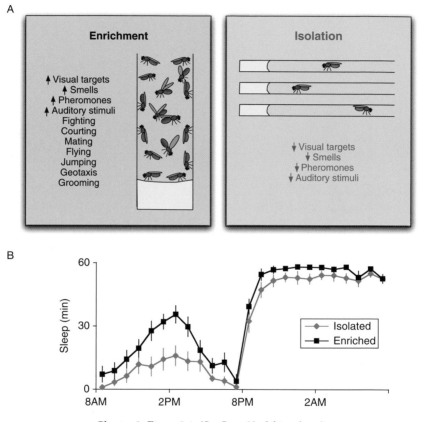

Chapter 3, Figure 3.1 (See Page 66 of this volume).

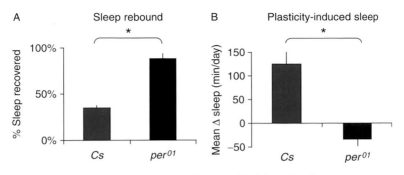

Chapter 3, Figure 3.2 (See Page 67 of this volume).

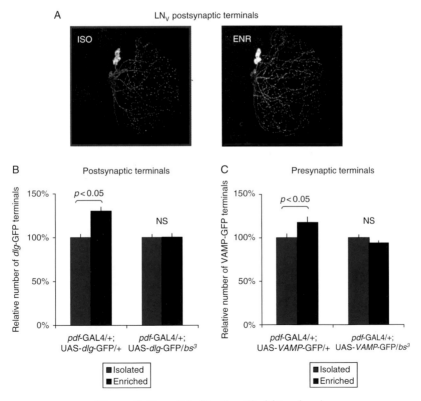

Chapter 3, Figure 3.3 (See Page 69 of this volume).

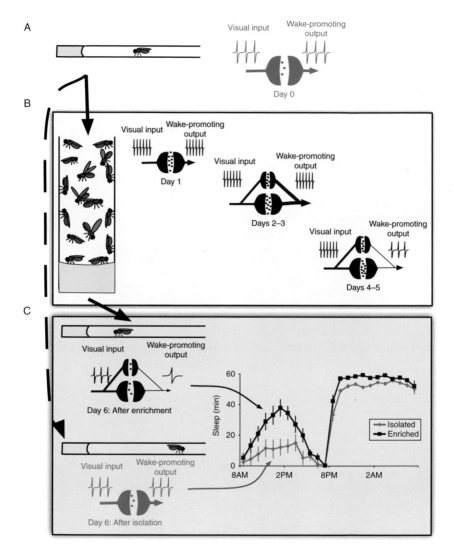

Chapter 3, Figure 3.4 (See Page 73 of this volume).